Cane Creek Days

WARREN GILL

Illustrated by Lissa Gill
& Peggy Gill Foster

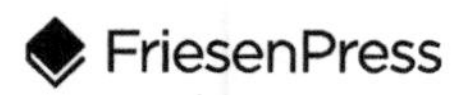

One Printers Way
Altona, MB R0G 0B0
Canada

www.friesenpress.com

First Edition — 2021

I dedicate this book to Lissa Gill, my wife of 43 years and my best friend.

Illustrator: Peggy Gill Foster

Illustrator: Lissa Gill

ISBN
978-1-03-910034-3 (Hardcover)
978-1-03-910033-6 (Paperback)
978-1-03-910035-0 (eBook)

1. BIOGRAPHY & AUTOBIOGRAPHY, PERSONAL MEMOIRS

Distributed to the trade by The Ingram Book Company

Introduction

(or why would anybody want to read this?)

I grew up on a story-book farm with cows and sheep, horses and a mule, occasional chickens, corn and tobacco and a beautiful little creek. The stories that follow happened to me and my family, and they happened along the Little Cane Creek. As creeks go, the Little Cane is probably not that special, but it has been important to the people who lived and raised families, crops and a large assortment of animals along its banks. Even casual visitors are drawn to its simple beauty and slippery personality.

Most people raised on farms have a head full of stories about the things that happened to them in their youth. Some of the tales make you wonder how any of them ever made it. Ask anyone raised on a farm if they ever had a close call where someone was badly hurt or almost killed. Or ask if times on the farm were ever tough. Be prepared to hear some good 'uns.

Most of the stories that follow are true or are at least based on truth. Not all the stories directly involve the creek, but most do. Some have been edited to keep me from getting in trouble with close relatives or friends.

I wrote this 'stuff' because my daughter, Greer, told me I should (and it is more fun to write than my usual technical papers). The real question is: Why should anybody read this? My family should read it in self-defense and to see if my memory is accurate. Others

- those who know me - should probably read it for the same reason. For those who read it who don't know me as well, I hope there are a few entertaining stories about a time largely passed.

The Cane Creek and Us as Children

A stream is like a magnet, drawing those who would seek the cooling gift of water and restful place of beauty. Big streams, if big enough, are called rivers. Little streams, springing from the ground, are the origins of running water and are called brooks. Creeks are between brooks and rivers. Someone once told me that the official definition of a creek is running water less than a hundred miles in length. Creeks are the arteries of the land. They bring the land to life.

Creeks are life. They have life within them and along their edges. They are the bottom of a valley, so all that walk and crawl the land, sooner or later, must cope with the creek that builds the shape and character of the valley floor.

People have many ways of refining and reshaping their world. Creeks are not immune from human intervention. Many words describe what people do to creeks: fjord, cross, preserve, pollute, seine, bridge; but this is more the story of what creeks do to people. Specifically, this is the story of the Little Cane Creek that starts near the crest of the Chestnut Ridge where Bedford, Moore and Lincoln counties meet and is fed by the Hannah's Gap Branch and the Barham Branch and countless small springs and brooks flow until they grow and merge into a full-fledged creek

- Cane Creek. This is the story of a section of that creek that crosses the Gill Farm.

The Little Cane Creek

For several years, in the late fifties and early sixties, school day mornings found three children walking the graveled drive along the Cane Creek to the stoutly built wooden stock gap across from a young mulberry tree. The oldest kid was me, Warren. The youngest was Gloria, and the middle one was Alan. We would wait on the side of the road until Mr. Bill Fullerton's school bus stopped for us.

On nice days, if there was time, we would be drawn to the creek. The creek magnet worked on each of us in different ways. My brother had a need to throw a rock into the water. My sister sought the fossilized fauna (crinoids) we called Indian money. Me, I needed a snake.

The best place for all our needs was a gravel bar which had grown in the turn of the creek at the bottom of the hill below our home. On one end of the gravel bar, the creek bank was solid bedrock, covered with moss and mud in places and bare limestone rock in shades of gray along most of the stretch that ended in our favorite swimming hole. Creek willows had grown in places on the edge of the gravel bar, adding stability and encouraging other plant life, like the pungent patch of spearmint near the willows or, further up the stream, green swatches of stubborn Bermudagrass growing as thick as a carpet.

Warren, Gloria and Alan on Alan's birthday.

Across the creek grows an ancient elm, hollow and missing limbs, leaning into the water and most certainly doomed to eventually be swept away, but stubbornly holding on in the belief that when it gave over, the water would have won a major battle in its eternal war with the bottom land creek bank.

We called the commonest snakes of the creek water moccasins, but they are simply water snakes, with no venom and flat-hued, brown to gray bodies which allow them to blend beautifully with the water smoothed rocks, gravel and mud of the creek bank. Sometimes the snakes are seen as they swim gracefully through still pools and other times they are found coiled in mud holes, but most often they hide under flat, gray rocks at the edge of the water. The occasional snake may hide a couple or three feet from the water but are most likely to be under those rocks which are part in and part out of the stream.

To find a snake, one has only to turn over rocks until a snake is revealed, usually tightly coiled and upset at being disturbed. There is often a crawdad or two and an assortment of many-legged bugs and worms to be found in the mini-ecosystem under the rock, but the eye is drawn to the snake, usually less than a foot long, but sometimes as much as two feet, or more. The little ones were most sought after because they could be put into jeans pockets for a quiet ride on the school bus and eventual fruitful commerce in the society of youth in the Petersburg Elementary School.

There are several ways to catch a snake, a stick with a forked head being the most common along the Cane Creek, but bare boy hands are also quite acceptable for the littler fellas. The six to eight-inch babies, plentiful in the spring, bit as hard as they could but could only barely break skin and soon gave up their struggle. In defeat, they sulked in resignation. The vanquished snake could coil in the palm of my hands for close inspection.

We'd sometimes pet them, but snakes express no enjoyment of such shows of affection (snakes don't purr). It might've occurred once or twice to try to use the snake to frighten Gloria, but she refused to reveal any girlish fear of snakes and the threat of the snake would yield nothing but little girl disdain, or even a sharp kick on my shins if she determined there was a need for older sibling punishment.

Usually, it was Gloria who saw the bus coming and told us to get going. The snake was carefully secreted in my front jean's pocket and we'd hurry to the bus. Mr. Bill Fullerton was the driver and, as nice as he was, he'd never allow a snake on the bus if he knew about it, so I'd be sure to hold my book satchel low as if I thought there was a chance that he could see the little hidden snake. Usually, there was no reason to say anything about the snake to anyone on the bus. They were mostly older kids and had their own worries and interests that didn't include snakes in boys' pockets.

I had a nice business going, selling snakes for a nickel apiece to the town boys who were as fascinated by snakes as I, but who didn't have ready access to the wonderful resources of the Cane Creek to yield wriggly bounty into their grubby hands. The country boys in my circle weren't as impressed with snakes and wouldn't give up their hard-won nickels for one, but the town boys always had more nickels anyway and were glad to part with them for snakes.

Mr. Campbell meets a snake

In the sixth grade, the snake business took a downturn when I decided to share my love of water snakes with Mr. Campbell, the kindest and gentlest of men. He was the minister of the Petersburg Cumberland Presbyterian Church and a teacher who combined teaching skill with an unusual willingness to serve others. Perhaps it was my admiration for Mr. Campbell that led to my actions on that spring day. I wished him no harm, in fact only wanted to share the joy of my find, but his reaction was unexpected.

I'd made my usual foray to the creek and had a particularly nice specimen in my possession; close to a foot in length, longer than I normally brought to school. Mr. Campbell was working on something at his desk and we were supposed to be studying, but I had to show him my treasure. I took my snake out, held it behind its head and walked up to the desk.

"Mr. Campbell?" was all I said as I extended my serpent wrapped hand to within inches of his face. I was proud of my treasure and wanted to share it with my favorite teacher. He looked up and stared at the small snake. I think, in the long moment that followed, that I began to understand the hypnotic effect that snakes have on their prey. Mr. Campbell was frozen. The man was literally locked into place by a tiny snake, but only for a moment. He later claimed that he thought at first that it was a fake. I admit his subsequent actions were consistent with that theory. He didn't move a muscle until the little snake flicked its forked tongue. A panicky look appeared on his face, and, in an explosive reaction, Mr. Campbell tried to jump directly backward. That might have worked if he'd been standing, but his backward leap from a sitting position resulted in a rolling backward tumble that landed him in a tangle with his chair in the corner of the room. His panic wasn't lessened by his ignominious rollover. He leapt to his feet and, while keeping widened eyes on the object of his fear, he quickly edged his way along the wall until he could break for the door.

It had all happened so quickly that it took a moment for the event to sink in on the rest of the students. One or two had watched the whole thing and theirs were the voices of knowledge that I could hear best among the buzz of whispers that was growing louder with every moment that Mr. Campbell was absent. I was worried. I'd meant no harm to Mr. Campbell. I loved him as much as any teacher I'd ever had, and I was afraid that this kind man would be angry.

I was getting ready to return to my seat when Mr. Campbell returned, but he stopped at the door. His first action was to stare at the room to hush the buzz. That done, he turned his attention to me. "Find something to put that snake in, Warren." That was all he said before he disappeared.

I quickly found a jar and put the snake inside and put the top on. When Mr. Campbell returned, he told me to take the jar and find the janitor and ask him to put holes in the top, so the snake could breathe. I did exactly that, except I added a stick and some gravel and leaves to make the snake more comfortable in its glass prison. When I returned to class Mr. Campbell eagerly took possession of the imprisoned snake. He was almost exuberant in showing the snake to the other teachers after it was safely locked into the jar. He took delight in the little snake in its new home and was curiously reluctant to return it to me with advice that it would be best if the snake could be returned to the Cane Creek to live out its natural days.

Cane Creek History

The Cane Creek has been attracting people for a long time. Arrow heads are abundant in the fields along the creek, which probably indicates that the hunting was good long before white men settled the area. There is, however, no evidence that Indians built permanent villages along the Cane. My wife, Lissa's theory is that, if we could we go back and ask the Indians why they didn't live in Cane Creek Valley, they would tell us that the rich soil, abundant streams, a rich diversity of plants and the temperate climate made for a long and prolific growing season. The plants made a good place for deer and rabbits and bison and bears, but trees and weeds also filled the air with pollen. Such an air-filling slurry of allergens makes the area a place of dread for anyone with allergic sensitivities who has intelligence enough to connect sickness with location. Indians probably had a name for the Cane Creek Valley or perhaps the entire great basin that meant `Place-to-hunt-where-sneezing-happens.' The Indians liked hunting here but were too smart to want to live here.

Joseph Greer was evidently not given to allergies or he was too stubborn to admit that something he couldn't see would make rich farmland an undesirable place to homestead. Maybe the fact that he owned thousands of acres in the area gave him extra incentive. He'd been given the land grant because he was the King's Mountain Messenger. He had been given the assignment to deliver the good news about the Battle of King's Mountain to Congress. He took off in a hurry on a horse but he was a big man (over seven feet by some accounts) and the horse died along the way. He walked some until he could find another horse. Once, he hid from Indians in a hollow log (the Indians who were seeking to remove his large scalp were said to have rested their haunches on the self-same log). This adventure was a big deal at the time and made him a hero.

When he got the land grant, he moved himself and his whole family to a lovely and productive spot on the Cane Creek and built a house. To be exact, his house was located on the Little Cane Creek a couple of miles up-stream from "The Forks" south of Petersburg where the Little Cane joins with the Cane as it winds its way toward the Elk River. The Cane or Big Cane is formed from the joining of the Middle Cane and West Cane with Sander's Creek just north of Petersburg. It isn't clear who else was around when Joseph Greer showed up, but it is known that the community of Petersburg had soon grown into a thriving rural community with a flour mill, a town square, a jail and several churches. Eventually there were said to be fifty businesses in this town that was formed near the beginning of the Cane Creek.

Joseph Greer's home is gone, but the house built by his son, Jeff, and a crew of slaves sometime in the 1850's, is very much still there on the Little Cane. Jeff Greer's daughter, Kate, married Hugh C. Moore and they lived on and farmed the land until it was left to their son, Allen Moore. My Grandfather (Papa Gill), Warren Gill, Sr., bought it during World War II in 1943. Papa Gill lived in the Moore House, until his mother (Mary Gill) died. At that time, he and Mama Gill moved into the Warren house which had been built by his Grandfather at the same time as the Moore house. Papa Gill sold the old Moore farm to my parents, William and Carolyn, and they moved into the house shortly after they married on January 1, 1950. I was born on December 20, 1950.

The house was not in good shape when Mama and Daddy moved in. There was no indoor toilet. The plaster walls were falling. The high ceilings (almost fourteen feet) were interesting to see but they made the drafty old house almost impossible to heat, at least in the early years. The conversion of this old house into something which is attractive and comfortable took dedication, money and years of hard work

Work is the operative word for the accomplishments along the Cane Creek. The creek itself built the richness of the land but raising livestock and crops on the land took a degree of toil which will never be truly appreciated by those who followed. In a perfect world, hard work yields the desired results. In the world along the Cane Creek, hard work sometimes yielded rewards, sometimes heartbreak, and often the results were not those intended, but often led people to make new (sometimes better) choices.

Bad Dogs (or whatever happened to all those sheep)

The night was perfect for hunting. It was growing dark in the early evening, but there was a full moon rising and the warmth of the day was fading. The running would be fast. The prey would be easy to see and a delight to torment. The dogs knew by instinct that this was the kind of night that their blood yearned for. One, the oldest and the leader of the pack, was half German Shepherd; another had the blood of hounds and the rest were of mixed ancestry, but all had the thrill of the hunt in their genes. They were town dogs, but they had their freedom and felt joy in escaping into the countryside. They understood the land and knew how to cross fields and crawl under fences and how to avoid the farmhouses that had a man with a big shotgun who hated dogs.

Most of all the dogs understood that following the Cane would take them to adventure and prey. When the lead dog's howl signaled that tonight was a good night for a hunt, it was a rallying call. They knew where to meet because they'd done it before. They knew the best time was after the moon was high enough to give them just enough light for all-out running.

Maybe the target would be a stray cat, or some goats. Cattle were no good as prey because of their size and strength and they could aim deadly kicks with uncanny accuracy. Cows took the threat of stray dogs seriously and responded with intent to do harm. It was much the same with horses, except horses are less predictable, and as likely to do damage in panic as in anger.

The best prey are sheep. Sheep are exactly the right size for dogs to hunt and chase. Sheep run away but are not fast enough to outrun dogs. One or two might, if cornered, try to fight, but they are no match for even two dogs, much less the five that were gathered on this Spring night.

Happy ewe and lamb (drawing by Lissa Gill)

This pack of town dogs are not hunting for food. All left home with full bellies. They wouldn't eat a single bite of their kill. Their joy is the thrill of the hunt. They lust for the taste of blood, and there is blood aplenty, but there is also an abundance of meat. These dogs are too excited to tear through wool and skin to the meat below. Why cease the hunt to tear into warm flesh when there are other sheep to chase and kill?

The game warden could instantly tell the difference between a dog kill and a coyote kill. Coyotes attack only one animal at a time, and they go for the throat for a quick kill. Coyotes take lambs when they can and rarely try for a big ram or ewe.

Dog kills are covered with random bites, as if the dogs had played with the sheep before finally torturing it to death. Dogs kill numerous animals and damage more than they kill. Blood is everywhere. Legs are broken. Wooly skin hangs in ragged tatters. Flies are drawn to the blood and exposed flesh and will be feeding on blood and laying eggs on flesh within a few hours after an attack. Fly larvae – maggots - will soon be squirming and growing in the wounds unless they are treated.

The sheep that survive a dog attack are shattered. Sheep, by definition, are flock animals. They are more devoted to being a part of a group than any other farm animal. When they are attacked by coyotes, and lose one or two animals, the flock absorbs the hurt and goes on. When dogs attack, the entire flock psyche is damaged by the torture, often irreparably.

Distressed Ram
(drawing by Lissa Gill)

Sheep farmers know also that the owners of the dogs almost never believe that their gentle house pet could cause such devastation. If you see their dog around your sheep and tell the dog owner, he'll get mad at you. Dogs are part of the family. You don't accuse a family member of being a sheep killer. The best course is simply to kill a suspect dog and hide the body. If the owner comes looking, express ignorance, even sympathy, but no guilt, not if you want to keep a friend.

Five dogs ran in the fields along the Cane Creek. They knew where the best places were to cross the Cane and its tributary streams. After they crossed Scott Branch, the lead dog, the German Shepherd, smelled sheep. He cut northeast for a hundred yards, then headed to a shallow crossing near a draw which would lead to where the sheep were pastured.

The creek was cold and one of the dogs, a young half beagle, hesitated for a moment, but a glance told him that his buddies weren't waiting. His pack instinct overcame his fear of the rapidly flowing stream. He crossed and soon caught up.

With whatever control he could wield, the leader kept the pack moving as quietly as town dogs could move. The sheep, some ninety or so ewes, had been shorn only days before. They would be faster without their wool, but they were still no match for the dogs. The German Shepherd was four years old and experienced. A couple of the other dogs were almost as canny and instinctively spread themselves for their attack. The young beagle, still trying to catch up, was on his first hunt. He didn't understand strategy, but he understood that sheep were prey. With the throaty bark that came from his beagle mom's lineage, he exuberantly threw himself into the field, where the flock quickly saw him.

Flock mentality is amazing. Whether resting or grazing, several of the sheep will be looking around, watching for predators. When is threat is spied, all the sheep reacted simultaneously. The half beagle sloppily ran one way then another as the flock reacted with group intelligence, not only with organized escape from the silly young predator but also by putting the strongest ewes in place to respond to the stupid pup.

The old German Shepherd had been around for a while, but the half beagle's loud, playful, sloppy attack was the stupidest thing he'd ever seen a dog do. Amazingly, the other dogs showed signs that they wanted to join the beagle, but the German Shepherd's throaty growl kept the other dogs from joining the chase. The old dog knew that the sheep would quickly reveal their escape path, and he would react accordingly.

When the lead sheep made for a cow path leading toward their home barn, the German Shepherd knew he had 'em. There was a brook they had to cross, and the crossing place was narrow. The flock would not be able to cross at speed but would have to slow at the bottle neck of the crossing. The Shepherd led his pack into position as the lead sheep crossed the brook. A couple of old ewes had broken from the back of the flock to stamp and threaten the beagle and had succeeding in confusing and cowing him. He had thought that sheep were cowardly, yet here were a couple that were trying to beat him up. Suddenly it occurred to him that he was by himself and that he'd maybe screwed up by not waiting for the pack.

That was when the rest of the dogs tore into the flock. The German shepherd attacked from the stream so that the body of the flock would be turned from their escape route. The rest of the dogs hit randomly. Flock wisdom was forgotten as each sheep scrambled to escape. The dogs each found their own favorite strategy for torturing their prey. The German shepherd tore into a huge black-faced ewe and rode her to the ground. His jaws found her neck and his teeth opened a huge gash that included enough of her jugular that she was soon to die. By the time she was dead, though, the big dog had left her in favor of her lamb. Although the lamb weighed twenty-five pounds the German Shepherd tossed it like a doll.

One of the dogs found joy in chasing a sheep until they could bite into its haunches and tear loose a flap of skin. Another of the dogs liked little lambs and killed at least five. The little beagle simply chased and chased and chased. He never bit anything but killed at least two by running them into the Cane Creek.

The orgy of death and destruction was halted with a shotgun blast. The little beagle rolled a couple of times under the weight of double-ought buckshot. He was instantly dead. So was the hound. After two quick shots and two instant kills, there was a pause while Daddy reloaded. The German Shepherd knew he had only a moment to escape and he didn't pause. He leapt from the yearling ewe he was chewing on and ran for the closest cover, a briar patch near the stream. He had disappeared into the growth and dropped to his belly to crawl into the water as the buck shot tore over his head.

A skinny cur-dog took the next shot into his rear. His death was slow. An old yellow part Lab, the only one besides the German Shepherd with a collar, had better luck. He got hit with the buck shot between the eyes. He was dead before he fell.

The final count was twenty-eight sheep dead, four dogs killed and a flock ruined beyond repair. Daddy was out of the sheep business. They just weren't worth it.

*****

This story, in one form or another, was repeated along the Cane Creek and countless others like it across Tennessee. The sheep industry was large and viable before and during World War II. In 1950, there were almost a half million sheep in the state. By 2000, there were less than fifteen thousand. Once, sheep were regarded as a reliable source of badly needed cash. They gave both wool and lambs, and the money came in the spring, which was a handy time because most other farm crops were sold in the fall. By the late eighties most of the sheep that remained were in small hobby flocks, club lamb flocks, with a few 'die-hard' serious sheep managers who had found niche markets. Many factors led to the loss of the sheep industry. Most farmers mention dogs as the main reason for getting out of the sheep business, but several other factors are involved. Foot rot, which was difficult to control during the fifties and sixties, is often mentioned. Low prices didn't help. Neither did the ever-widening use of Kentucky 31 Fescue as the base forage on farms. Cows do better on fescue than sheep.

The most important factor in the loss of the sheep industry was the fact that farmers got jobs in town. They no longer had time to fool with sheep, and sheep need attention. Sheep are probably the most thoroughly domesticated farm species. They have evolved to the point that they cannot survive on their on - they need us. We used to need them, but now we don't. We don't eat much lamb and wool is grown more efficiently in other places in the world. Also, wool has increasingly been replaced by synthetics.

Sheep are often called stupid. Taken individually, they are a little dim-witted, but that is probably a by-product of an evolutionary surrendering to the protection of the flock. As a flock creature, they have exactly the correct amount of intelligence, particularly since the domesticated sheep should be viewed in the context of having a shepherd to protect them. Wild sheep are not regarded as stupid. It is the modern domesticated sheep that must have human protection to survive.

Instead of focusing on the stupidity of sheep, consider how they have almost exactly the correct amount of intellect to fit their environment. After all, how smart do you need to be to graze, eat hay and chew a cud? A ewe only has to eat, get bred and have a couple of lambs a year. They are an almost totally predictable animal. You move one way, they respond in an exactly predictable way. You put 'em on their butt, they sit still. If you don't understand sheep, and can't get them to do what you want, chances are you will think them stupid. People who understand sheep don't usually call them stupid. Maybe the people who call sheep stupid should reconsider their own level of intelligence, or at least their understanding of animal behavior.

Bridges Across the Cane

Crossing a creek is not usually a major problem in life. The Little Cane is not a large or treacherous stream, so it has many places where it can be forded with ease, but that kind of crossing was not sufficiently reliable for motor vehicle traffic. Hence bridges.

The bridge over the Little Cane on the east end of the Gill farm was first built early in the twentieth century and served well through most of that period. It was constructed of concrete and oak with steel side rails. It was an important part of what made travel on Gingerbread Road possible during practically all weather, the only exception being when a huge rain brought the Little Cane out of its bank and around the bridge.

View of Little Cane Creek from Gingerbread Road bridge

When cars crossed the bridge, the noise of the boards rattling against steel I-beams carried at least a mile. There was no way to sneak over that bridge without being heard, not that there was any reason for sneaking up Gingerbread Road.

The Wilsons lived nearest the bridge next to the large pond where Cold Spring Hollow Road joined Gingerbread Road. Braden and Annie Lee moved there in the late fifties with Annie Lee's mother, Big Mama McDaniel. Their daughter, Bradeen, was a frequent visitor and she subsequently added her adopted daughter, Barbara, to the mix. These fine and lively people came to the country after Braden and Annie Lee retired from a life of running a successful grocery store in Lewisburg.

The Wilsons were truly interested in the lives of the people who lived and visited along the Little Cane and one of the ways they kept track was by listening to the sound of people driving across the bridge. Mrs. Wilson claimed she could tell from the sound alone who was crossing, and she probably could. She said that Granny Collier flew across the bridge, so the sound was brief. Papa Jepp's old truck had plenty of its own sounds and he tended to drive more deliberately (slowly), so it was not surprising she could tell when he crossed. It was a little aggravating when she told anyone who would listen that the Gill boys drove across too fast, too often and too late at night.

Mrs. Wilson loved to go fishing, and one of her favorite places to fish was from the bridge. Maybe the hours she spent on the bridge, fishing pole resting on the steel side rails, with cars and trucks and the occasional bicycle riding by, were how she became so adept at reading the sounds of the bridge. Rare was the traveler who didn't stop to pass the time with Mrs. Wilson, and many were standing beside her when she hauled in a sun perch or blue gill. Many of the fish were too small to eat, but some of them were roughly the size of a man's hand and worthy of a place in her frying pan.

There was a smaller bridge (later replaced by a culvert) at the foot of a long, steep hill within a hundred feet of the big bridge. The study of the sounds that was such an important part of Mrs. Wilson's tenure was incomplete without understanding that the little bridge added its own sound. If you drove fast enough down the Gingerbread Road's big hill and carried that speed across the small bridge, there was only a muffled whiffle of sound, followed in a couple of seconds by the rattling banging of the big bridge. On a summer day, when the dirt road was dry, all this noise was accompanied by a cloud of dust that blew up around and behind whatever vehicle was traveling by.

When we were kids, a great way to pass time was to ride or walk a bicycle up the big hill and coast down. You could fly, even to the point of hearing the wind blow across your ears, and there was an element of danger to make it exciting. Gingerbread Road was, at that time, graveled with creek gravel over a rocky, dirt base. The result of years of auto/truck/tractor traffic, with occasional, pre-election attention by the road department, was a base of hard packed dirt with two dusty tracks where tires ran and a graveled roadside with rocks ranging in size from tiny specks to as large as a grape fruit. The larger ones tended to work into the ditch, but fist-sized rocks were common and could easily find a place in the dirt track. These loose stones were the ones that were most likely to throw a speeding bike into chaos, but there were also exposed parts of buried limestone boulders, smoothed by car tires, but fully capable of jarring and upsetting an unwary cyclist.

The essence of the game was to get to the top of the big hill and pedal down until there was enough speed to keep you going. Then, you stopped pedaling and let gravity take you down. The idea was to see how far you could coast. It was not difficult at all to coast across the little bridge, but it took skill and nerve to get enough speed to make it to the big bridge. The skill part was related to avoiding rocks and loose gravel. The nerve part was pedaling to gather speed while avoiding the use of brakes.

I had a narrow-tired bike which gave me an advantage in our contests, but Alan was utterly without fear as he pedaled down the steep hill. One of our rules was that you had to stop pedaling at the post on the north side of where the Cold Spring Road joined the Gingerbread Road. My strategy of building maximum coasting speed with minimum braking got me almost to the bridge, a difficult mark.

Alan's strategy, to my amazement, was breakneck speed coming down the big hill, attained by nonstop pedaling. Think about it. If you are pedaling a bike, there is no strategic braking going on. There is no braking at all. There is only more speed.

If skill is involved, Alan is the best at it. He mastered the hula hoop quicker than me. He was so good at jumping a pogo stick that he could keep it going for hours. He'd learned how to ride a bike in practically no time and by the time he was pedaling down the hill in our contest, he was a true master of all bike-related skills.

His bike was red, but that was of little importance as his speed built on that memorable day. His unshirted back was parallel to the bike as he bent himself low over the handlebar to cut wind resistance. His legs, made strong by countless hours of bicycling and baseball, were tireless as they pumped harder and harder, like pistons driving a racing engine. The tires spun faster than the human eye could follow. He strained every muscle in his body to squeeze more and more speed from the hurtling bike.

I watched first in admiration of my brother's nerve, then in dread because I knew he'd probably win this contest, and he'd doubtless rub it in. He'd probably tell everyone. Then, as his speed grew beyond anything I'd ever seen before, I became afraid. This was just too much! I even tried, with big brother bossiness tinged by panic, to scream at him to slow down, but there wasn't a chance he heard me because that was the same moment when it happened. Maybe it was loose gravel or perhaps his speed simply grew too great for the bike to maintain stability.

It started with wobble in the front wheel followed, in less than a blink, by a bike sideways and down and a boy sliding through gravel on his front and side. It seemed like he slid twenty feet, but it was probably less. Whatever the measure of the distance of the slide, the skin of the human boy is not designed for such abuse. By the time his slide had ceased, there was more skin left in the dust and gravel than remained on his bleeding belly.

There was dirt and gravel embedded in the skin over his stomach and side, and some in a couple of scrapes along one of his legs. There were deep, bleeding cuts and shallow scrapes and some bruises to add a little more color.

I'll never forget his bravery as he accepted his fate. He was more stoic about it than I was, and I was only a witness. I don't remember if he cried or not, but he sure scared the hell out of me. Maybe the scariest part of all was the long walk to the house. He walked stiffly and he started slowly, but got in more of a hurry as we approached the house. Mama was in the kitchen, but soon answered my cries by meeting us on the porch. The look on her face was a mixture of fear and concern, maybe puzzlement, because this one was so much more than our usual scrape or cut, but somehow we got through it, and it became an entertaining story to tell my friends at school.

A John Deere Meets the Cane Creek

(and Warren learns a tractor driving lesson)

John Deere tractors are built for power and reliability, but they can force one to face reality in unexpected ways. We had two of them shared in some family arrangement between Papa Gill, Daddy and Uncle Edward. One was an M model 'poppin' John' that was started by taking the steering wheel off and using it to turn the fly wheel until a mechanical mix of whuffing valves and exploding pistons signaled that the powerful machine was coming to life. Big John was what we called this awesome machine which, to my young self, was powerful enough to move mountains.

The smaller tractor, Little John, was handier. For one thing, it had a starter with a real ignition switch. It was also easier to handle so was the natural choice for a boy to learn tractoring skills. Little John had only three forward gears and I'd conquered two of them by doing odd jobs for Daddy. I'd pulled a trailer filled with brush, and I'd run the disk. I had even driven the tractor in low gear while Daddy and others loaded hay.

The tractor wasn't so hard! I had it down, but confidence has a way of dissolving in the face of reality. One day, apparently without thinking, Daddy told me to bring Little John to the 25-acre field where he was getting ready to cut hay. It occurred to me that this would be the first time I'd driven on the road. I started to ask for advice but remembered how thoroughly I'd mastered the art of tractor driving, and decided driving on the road wouldn't be a big deal. People did it every day.

Trouble is, I'd never driven a tractor in high gear. Tractors are built for pulling power and not for speed. Cars are heavy and tractors are heavy, but they are heavy in different places. The wheels are different, and the gears are different. Twenty miles an hour is creeping in a car but is fast on a tractor. Road bumps that a car absorbs with ease are dangerous obstacles to a tractor. Turning at fifteen miles an hour is nothing for a car but can be life-threatening on a tricycle-wheeled tractor.

Daddy had built a stock gap from the straightest Bodock poles he could find. He had placed the gap in the driveway just where it met the Gingerbread Road less than twenty feet from the little bridge. It was too close and would later be moved, but in those days, drivers had to start turning the second they crossed the stock gap if there was any hope of making the turn onto the bridge. It was such a tight turn that the school bus couldn't make it until Daddy lengthened the stock gap. Then, possibly because the lengthened gap was weaker, the bus wheel fell through it one morning and Mr. Fullerton decided to quit trying to cross the gap at all.

None of this was in my mind that morning Daddy told me to drive the tractor on the road. All I could think about was that I was finally going to get to drive in high gear. I couldn't wait. I would be driving faster than I'd ever dreamed possible.

In my excitement I made two errors. The first was in ignorance. I knew that tractors had two brake pedals, one for each big rear wheel. I knew that you pressed the left pedal to operate the left brake and the right pedal to stop the right tire, and I even knew you could mash both pedals to operate both brakes. What I didn't know was that there was a little latch that you could (and should) flip to fasten the brake pedals together when driving at road speeds because you never, ever wanted to mash only one pedal when the tractor is going fast.

The next mistake was to set the throttle too high. This was a product of my over-confidence and my desire to go as fast as possible as soon as possible. I put the tractor into high gear on our gravel driveway some ways before I got to the stock gap. The tractor started going and it felt great. Then I got to the gap and shot across like a dream, and I started making the immediate left turn right on schedule.

In my young life I'd never been in a situation like that before, and it contained a valuable lesson about thinking things through before getting into the middle of them. The tractor was going too fast to make the turn. The front wheels slid in loose gravel and were skidding toward the right edge of the bridge with no possibility of stabilizing in time to make the turn. With only inches to spare and less than a second to act, my young feet instinctively found the brake and I pushed with all the power that my panicked legs could muster.

The brakes caught and the threat of driving off the east side of the bridge was instantly replaced by a lurching turn in the opposite direction. I'd only managed to push the left brake and the right tire was left free to hurl the tractor off the west side of the bridge, which was exactly what happened.

There followed a moment of which I have no remembrance. I don't remember it now, and I didn't remember it then. I remember not remembering it. I couldn't explain to anyone, even minutes after it happened, exactly how the tractor wound up in the creek. The next thing I knew was that I was clinging to honeysuckle vines on the side of the deep ditch and the tractor was below me, in the middle of the stream, still moving. It had somehow flipped into the ditch and landed right-side up, except it was now churning its way through shallow water, slippery creek rocks and mud back toward the bridge.

Daddy appeared on the bridge. He barely glanced at me, saw that I was alive and instantly decided that the most important job for the moment was to stop the tractor. He didn't hesitate but leapt from the bridge and landed on the narrow top of the engine housing. The John Deere was lurching and rocking as it fought its way upstream. He walked the top front of the tractor like a tight-rope walker and, in the blink of an eye, had used the steering column as a hand hold for twisting his body into the tractor seat.

The second he sat down and started to take control of the tractor, the engine seemed to recognize that there was no more point to the struggle, and it died. About then, I came to my senses and decided that clinging to the vines was serving no purpose. I pulled myself up on the bank and, to my surprise, looked down as Daddy started the tractor. He skillfully backed the tractor down the stream the thirty or so yards to where it joined the Little Cane Creek. I feared I'd probably ruined the tractor forever, so was immensely relieved as I watched him drive Little John up the bank onto dry land. That was when we noticed that one of the big wheels was running crooked. The steel had torn during the fall and would require removal and taking to a welder before Little John would be serviceable again.

Pete Phillip's bridge

As the years passed, the old bridge across the Little Cane became ricketier and more dangerous. Workers could replace the oak boards, but the concrete abutments were aging and eroding. Steel girders were pitted and rusty. The little creek had spent many years wearing away the very rocks upon which the bridge was built.

At the same time, traffic had changed, even on this dusty road. There was more traffic, and the size of the vehicles had changed. For many years, milk was picked up by simple box trucks, but modern dairying required large tractor trailer trucks with huge bulk milk tanks. Tractors and implements grew larger. Automobiles became larger and faster. The old bridge was not designed for such vehicles.

School buses were also larger. The kids who lived on the Gingerbread Road and up the Cold Spring Hollow were among the last to be picked up in the morning and the first to get off in the afternoon. The bus was usually almost full when it crossed the old, dilapidated bridge. It was a dangerous situation.

Contacting the county Roads Superintendent did no good, except to find out that it was the Tennessee Department of Transportation that took care of bridge matters. We wrote our congressman, Pete Phillips, a man who had been a state Representative for so long that he had seniority over sedimentary rocks. To appearances, he fit every stereotype of a backslapping, big talking southern politician, but, in truth, was a dedicated and energetic public servant who knew all the strings to pull to get things done for his constituents.

He called Daddy and suggested we make a trip to Nashville. Mama suggested we take pictures of the bridge, so I shot an entire roll of prints that showed every cracked and flaked piece of concrete, the rusted steel, and the old boards.

Daddy, Alan and I pulled out early one morning and got to Pete's office just as the doors opened. We only had to wait a moment before Pete came out and greeted us like we were his best friends in the world, but he wasted no time in getting us herded out the door and through several corridors, across the street and into the building that housed the Tennessee Department of Transportation (TDOT). While we were walking, he predicted the outcome of the meeting he'd set up with some low-level planning engineer. He said we would be treated with respect and would be told why the bridge couldn't be replaced at this time.

This was exactly what happened. The middle-aged bureaucrat who met with us was very nice, very professional, and armed with facts and figures about the ages and conditions of bridges across the state. He had an engineer's report on the little bridge that said it had been inspected and was safe. He also said it was due to be replaced in about ten years.

Pete asked him to look at our pictures and he gave them a polite glance, but he wasn't very interested. He clearly had said all he was prepared to say, and Pete led the way out of his office, never failing to remain upbeat and giving everyone a hearty greeting. We were disappointed.

As we were walking back to Pete's office, he said, "That was exactly what I expected. Right now, our host is filling in his boss about the visit. They hope they've heard the last of us, but if you want your bridge, we'll have to keep up the pressure. You will need to come back in a few weeks."

I'm not sure any of us thought there was much accomplished in the first visit, and I don't know if we thought it was worth it to make another, but Pete called, and we went. This time, we were met by a higher-ranking bureaucrat. He had the same facts and figures and was a little more accommodating. He seemed interested in the pictures, but still didn't encourage us in any way.

About a month later, Pete called again. "I think this one will do it," he said.

Alan couldn't make this trip, so it was just Daddy, Pete, and me. For the first time we were met by a team, led by the highest ranking TDOT official yet. They had reserved a meeting room for us and had obviously prepared for the meeting. There were several people in the meeting. One of the men, an engineer, was from Lincoln County and knew Mama. It turned out that his brother, Jim Jeans, had been the coach at 8th District School when she was teaching there. We spent some time in a friendly chat until we turned to discussing the bridge.

We got a more detailed explanation of how decisions were made regarding replacing bridges. They showed us some additional information about the inspection schedule of the bridge and congratulated us on our interest in working to improve the roads. Pete seemed to be extremely pleased that he had stirred up enough interest to bring all these TDOT people together in one room to discuss a bridge across the Little Cane Creek. He asked a lot of informed, intelligent questions and got detailed answers. During the meeting itself, however, we still didn't get any encouragement. All we got was a promise to look at the schedule to see if our bridge could be moved into a different priority.

As the meeting broke up, Pete seemed to be pleased. He shook everybody's hands and led the way out of the meeting room. While everyone else left, the man from Lincoln County came over to me, motioned for me to hold back. Then he started talking. I thought he wanted to discuss something about our home county, but I was wrong. He let everyone else leave the room, then, in a lowered voice so only I could hear it, said "You got your damned bridge." He said it almost in a growl, but the growl was soon followed by a smile.

I started to say something, but he continued, "Pete is unbelievable. He'll wear us out if we don't build it. You don't have to come back anymore. It will be built, probably within six months." He shook my hand, smiled again and left. By then, I had to rush to catch up with Pete and Daddy.

The bridge was started within three months.

Oddly enough, when Daddy tried to give Pete a donation for helping us, he wouldn't accept it. I think it is likely that Pete Phillips enjoyed winning a friendly battle with TDOT while he helped his constituents.

Fishing the Cane

Fish are tantalizing creatures. They swim in people's imaginations almost as well as they swim in water. Who hasn't dreamed of swimming like a fish? We emulate them when we try to swim ourselves, but our efforts are clumsy by comparison. They streak through the water almost without effort. They turn in the blink of an eye. They survive the coldest winters and even drought. Creek fish have been inhabiting the Cane since long before people came and will likely be there when people have been long gone.

The twists and turns of the Cane provide a complex series of differing watery homes. There are places where the water runs swiftly down an undulating, smooth-worn bed. The water sometimes runs in ways that allowed it to gouge long, small rivulets which are home to minnows. These are maybe two or three inches wide and about as deep, just the right size for children's hands to fit as they chase and trap the little gray-brown sucker minnows. The rock creek-bed is slippery with a layer of bottom-scum making it easy to slide up stream and down in the running water.

The slick rock makes standing tricky for young minnow hunters. Many a minnow has escaped young hands when the gleeful youngster slips and falls. Those falls often occur after a warning by some parent about falling in the creek. The scenario is predictable. Sunday afternoon after a big dinner, grownups want to relax and talk, and kids announce they are going to walk down to the creek. "Don't get wet. You don't have anything else to wear."

"We're just going to look."

"Don't stay long. We got to go see …"

The kids are gone by now. Within fifteen minutes, shoes and socks are off and britches are rolled up as one, then another, then the rest of the kids go wading. I suppose that somewhere, sometime, someone has gone wading and not fallen in, but it has never happened in the Cane. The bottom is too slippery for human kid feet to sustain balance. Inevitably, one kid falls. Then, maybe because the falling looks like fun, the rest go down.

The only way to prevent the falling-in-while-wading syndrome is to put swimsuits on before the kids go to the creek. Sometimes grownups are smart enough to think of this, but often they just let the kids fall in, maybe because it gives them an opportunity to glare at the miscreant kids as they come dripping up from the creek after an hour or two of 'accidently' falling over and over into the creek. Usually, a scramble results in finding some poorly fitting dry something to put on while the wet clothes are put in the dryer. It all seems to work out somehow.

Suckers

Both the Cane and the Little Cane are home to an abundant variety of fish. There are sun-perch and rock bass and bluegills but the fish to be found in most abundance is the sucker. Suckers are widely considered as 'rough' or 'trash' fish, of some kin to carp and buffalo, which have a bottom feeding habit which leaves them unimpressed by spinner bait or dangling worms. Most sport fishermen are glad that suckers are not attracted to their bait because they have been told that sucker flesh is too bony to eat.

Perhaps the hard lives of the people who had lived along the Cane and other similar creeks in middle Tennessee had forced them to be resourceful in finding food. The trashy suckers that most people ignore provided an opportunity for poor farmers to convert an available resource into free food. That men developed a variety of fun, even thrilling, methods to catch suckers is tribute to their cleverness, but the ways that women found to turn bony fish into delightful entrees is truly on the edge of culinary genius.

Catching Suckers

(Rocky Hole)

Among the ways that have been developed for catching suckers are seining (hard work, illegal and not as successful as it might seem it should be), dynamite (too loud and also highly discouraged by game wardens), grabbling by running hands under banks and rocks (primitive but often successful if you can get past the fear / thrill of putting your hand into the dark and slippery world of snakes and snapping turtles), gigging (great fun when suckers are on shoal) and snatching. Of these, probably the most fun is snatching.

A good snatching hole has a combination of several ingredients. It needs to be long enough to give the schools of suckers room to swim up and down stream feeding on the bottom and narrow enough that the strong cotton fishing lines can be thrown most of the way across and still be seen by the fisherman. The creek banks need to be high enough to allow the snatcher to look down into the hole or have climbable trees growing out over the creek to accomplish the same purpose. The water needs to run three or four feet deep over a graveled or solid rock bed with as few snags as possible.

The Rocky Hole had all these and more. For one thing, it was on Papa Jepp's farm, so it was handy to get to. The Little Cane ran the south border of his bottom land until it met and ran alongside the Chestnut Ridge Road for a mile or so, then branched off to split the Wilson and Gill Farms. The Rocky Hole started near the southeast corner of Papa Jepp's farm. To get there, you start by taking the long cane fishing poles from their hidey hole under the house. You check the large treble hooks to make sure the points are harp, but not too sharp. If the points are overly sharpened, they will pierce a scale and come off the fish. If they are just right, they slide across the scale and buriy in flesh and the fish is caught.

Rocky Hole – Papa Jepp's favorite fishing hole

From there, the next stop was to cross the road, then drive across the bottom pasture to a ditch that you had to approach at just the right angle to have an easy path avoiding deep ruts and the old sinkhole. We were close, but still had to stop at the gravel bar below Rocky Hole to fill a canvas bag with rocks for driving the fish. The rocks needed to be just the right size for throwing. Papa Jepp taught us well that the fish would simply ignore a rock that was too small, and large rocks filled your bag with too little ammunition for a sustained fish drive.

Next, we cross the creek. It was deep enough that the water would wet your britches if you didn't roll them above your knees. Papa Jepp had rubber boots that allowed him to wade with little trouble, if the water wasn't too deep. Many times, to hurry things along, he or Daddy would carry us across, with us tightly clinging to their back.

After crossing, we'd climb a gentle incline up the bank, following a path that had been worn by countless fishermen over the years. Some fifty feet or so up from the crossing the water started getting deep enough to start seeing fish. Many times, the first fish we'd see would be a hog sucker. Hog suckers aren't as large as white suckers and are more typically seen as a single fish basking in shallower water over a gravel bar. Their mottled brown coloration matches the gravel bar almost exactly, making them difficult to see unless they are on the move. Usually, the only time we'd fool with a hog sucker is when we couldn't find white suckers, but, if it was a decent size, sometimes Papa Jepp would throw out his line and quickly snag one.

White suckers were the goal. They travel in schools of fifteen to fifty fish and are a deep gray color on the top and shiny white underneath. Often the location of the school was betrayed when the searching eyes of the fisherman caught a glimpse of white belly as the fish swam by. Most often, however, the first look at their white belly came when the fish was snatched from the creek.

We often paused for a moment when we got to an ancient Cedar tree with partially exposed roots digging deep into the bank and forming a kind of ladder on which we'd stand for our first real look at the deep water of Rocky Hole. This was where Papa Jepp would put on his sunglasses. The invention of polarized sunglasses made sucker snatching an altogether different proposition. You can see beneath the surface far better with polarized sunglasses and the ability to see the fish as they cross the white cotton line is crucial to success in snatching.

It is worth mentioning that weather is important in sucker fishing. It must be a calm and sunny day. If it is either too windy or too cloudy, you cannot see the suckers. If it has rained enough to make the water dingy, you cannot see the suckers. If the water is too deep from flood waters, guess what - you can't see the suckers.

Papa Jepp had incredible vision. He'd almost always be the first to see the school of suckers, and he usually saw them as he leaned over the creek while holding a branch of the Cedar tree. The suckers were usually holding themselves within a shady spot by gently swimming against the current. After Papa Jepp found the suckers, he'd direct us to our places. The logistics were important and changed depending on who was with us and how experienced they were. Papa Jepp was the best at snatching and Daddy was almost as good. Uncle Jim wasn't bad, but not as good as Papa Jepp or Daddy. Mr. Bert Nichols sometimes went with us, but he was only a passable fisherman. Kids under twelve or so were simply rock throwers, but after we got strong enough, we were given occasional opportunities to try our hand.

The best place to fish from was an old elm that leaned out over the creek. It didn't have many branches which made it easy to fish from, but difficult to climb. To solve this problem, Papa Jepp nailed planks on the old tree so he could climb it. He usually took his place on that tree, threw his line into place across the middle of the creek, maybe a little to one side or the other, depending on his idea of where the fish would run and direct us to start driving the fish.

The first time up the creek, he'd want us to drive them slowly. "Don't stir 'em up, Warren. Throw way down behind 'em. Alan, you stand back for now, but be ready to throw toward the far bank." Even as a little boy, Alan's throws were accurate and could be relied on to be just right for turning the fish in just the right direction for crossing the snatch line.

The author demonstrates snatching using Papa Jepp's actual cane fishing pole.

Lead sinkers are placed some three or four feet up the line from the treble hook. Those few feet of line between hook and the weights are where the fish need to cross. The line is heavy and white and the fish can see it and will turn around if they have time to think about it, so they need to be going fast enough to not have time to pause and turn, but not so fast as to make snatching difficult. Papa Jepp could catch fish no matter how fast they were going, but a slow crossing let him pick out the biggest fish. Sometimes, he would even catch two at once if he had two hooks on the line.

Snatching isn't easy. It takes a combination of naturally good reflexes and years of experience to make a good snatcher. Strength is useful but is not nearly as important as timing and touch. Snatching too hard results in a miss or 'turning' the fish. Turning means that the hook slid under the fish and grabbed it enough to turn it over, revealing the white belly, but the fish isn't caught. Often, when the hook is inspected, there is a scale on the hook. An overly hard snatch drives the point of the hook into the scale.

A 'turn' sometimes occurred when the fish was over the sinker when you snatch. If the sinker turns the fish, it is unlikely that it'll be caught.

An extremely strong pull can tear through the fish. This not only lets the fish gets loose but can also result in a wounded fish swimming around dragging its guts from its belly. Sooner or later the wounded fish weakens, and it can often be snagged as it floats into the shallows.

Every snatching fisherman has misses and turns, but a good fisherman doesn't have as many. The best fishermen, like Papa Jepp, are almost always adept at other things that require good hand-eye coordination. Papa Jepp was an excellent baseball player and could throw with more accuracy than almost anybody. He was a crack shot. I've seen him shoot a chicken in the head with a twenty-two. Sounds easy? Try it. Better have more than one cartridge.

He had the 'touch' for fishing. He knew just the moment to pull and applied exactly the right pressure to the pole to send a ripple of energy down the line to make the hook slide under a scale near the front fin, often just behind the head, to slip the unsuspecting sucker from his school buddies, wriggling and fighting into the air and onto the ground. If Papa Jepp missed, his line didn't leap far from the water, so he rarely got tangled in overhanging branches as less skilled fishermen did. This allowed him to return his line to creek bottom within seconds, ready for another chance at impaling a good 'un.

Alan or I would hurry to remove the fish from the hook and check for line tangles and clear junk off the hook so he could quickly get the line back in the water. We'd usually string the fish and keep it with its unfortunate fellows in a little square pool formed by three boulders that had long ago broken from the creek bank, but whatever we did with the fish we had to do in a hurry because the school needed to be kept moving or they might make an escape under the creek bank or beneath large rocks. Sometimes, if the fishing was going well, we simply cast the fish back on the ground away from the bank where its flopping was unlikely to result in escape back into the water. Later, after the fishing had slowed, we'd go in search of the ones we'd thrown back and string them as we found them.

Believe it or not, snatching has its hassles, problems, and downright headaches. First, you gotta have a good pole. Finding a strong ten to twelve-foot cane pole isn't easy. Papa Jepp was always on the watch for a good one. He also spent countless hours taping and bracing old ones that he liked.

Another problem is getting hung up. Every fishing trip included at least one or two times when the hook and line got snagged in a tree limb. Often, the more you try to get it out, the more it gets tangled. If a kid could climb to the tangle, up we'd go. We always had a knife if we couldn't untangle it. We'd hack at branches and, usually, it would soon be free. In cases where the snag was too high to reach and too tangled to free, the decision had to be made to cut loose the hook. Papa Jepp hated to lose a good hook. For one thing, they cost money, and he didn't like to waste money. Also, he spent considerable time getting the hook to just the correct level of sharpness. A new hook was not 'broke in' enough. Eventually, though, the need to get back to fishing would force the decision to cut the line and reset the hook and sinkers.

Other mishaps included falling in - not too much of a problem unless it was cold. Cold legs on kids didn't bother Papa Jepp too much unless Mama or Granny showed up to check on us and found one of us shivering. Then he'd profess concern, but his mind was more on fishing.

Uncle Jim had an interesting problem one day. He had made a good, hard snatch that missed. His line whipped around a sizable elm branch and the hook came back and the exactly correctly sharpened hook buried itself into the meaty part of the palm of his hand. The expression on his face was one of puzzlement, like what just happened here and how could it have happened to me?

The branch had bent in the process, so was still exerting pressure on the line and on the hook in Uncle Jim's hand. The first step was obviously to relieve the pressure, which Uncle Jim tried to do by using his free hand to pull the line which bowed the branch but did not loosen the snagged line. He was on awkward footing and he slipped and let go of the line which let the branch pull it taught and bury the hook deeper in his hand.

By now, Uncle Jim's face was showing some level of distress. His hand was hurting bad, but he had to hold it out over the water to keep the line loose and relieve the pressure. He had no desire at all to repeat his attempt to shake the line loose. He needed help and got it. Daddy pulled a knife out and cut the line.

We all gathered around, looking with fascination at Uncle Jim's impaled hand. Remember that a hook is carefully designed to be easy to get in and very difficult to get out. We all looked to Papa Jepp for a solution and he had it. "Jim, we can't pull it out or it'll tear your hand up. We gotta push it through."

Uncle Jim wasn't too thrilled with that plan. "It'll hurt."

"Yeah, I reckon it will."

"Don't do to wait," said Papa Jepp. "But we'll have to go to the house to get some wire-cutters to cut off the large eye with the string still knotted in place. When the wire pliers were located, Papa Jepp grabbed Uncle Jim's hand and started pressing down to cut the metal. Fishhook metal is hard. Cutting it requires strength and struggle and the struggling was not fun for Jim.

By then, Uncle Jim was probably six-two and Papa Jepp maybe five- seven. Papa Jepp was strong, but Uncle Jim wasn't weak. Papa Jepp was straining to cut the hook and wasn't making any real progress but succeeded in finding a buried nerve or two. Uncle Jim retook possession of his hand.

"I'm going to the doctor," announced Uncle Jim.

"He'll do the same thing."

Uncle Jim was in college at the University of Tennessee in Knoxville, so he knew a thing or two. He said, "Yeah, but he'll give me some painkiller."

The fishing had been good that morning. The weather was perfect, and the water of the Rocky Hole was clear and at just the right depth. I could tell that Papa Jepp was torn. After all, Uncle Jim was his son and one of his favorite people in the world. But there was also fishing going on. Uncle Jim read his mind. "I can go by myself."

"I'll go," argued Papa Jepp, half-heartedly.

"No, I can drive just fine."

Daddy said, "It's his left hand, so he can shift all right. Maybe Warren can go with him."Daddy also wanted to get back to fishing.

"Y'all need Warren and Alan to drive fish. I can go by myself."

"Go on," said Papa Jepp. "Maybe you can get it fixed and get back in time to fish some more." Sure enough, he got back and joined us at Rocky Hole, but just watched and showed off his bandaged hand.

Seining

Uncle Edward preferred seining to snatching. Nets and seines may be regular gear in ocean fishing, but they aren't a common piece of equipment along the Cane Creek. Uncle Edward had a good seine and he kept it fixed to fish. It consisted of a long net, maybe ten feet, with a rope strung with oval shaped wooden floats along the top and a weighted rope along the bottom. Large poles on either end held the seine together and provided the handles that were needed to haul the seine through the water.

Uncle Edward's seine was large enough that it took two strong men to handle it. Creek seining, where there were plenty of obstructions, took quite a lot of work, and it helped to have a couple of boys following closely behind the seine to lift the seine over rocks and submerged logs.

Seining isn't legal. In fact, it is very illegal, especially if a bass or other game fish is captured and winds up in the fish bag. Maybe that's one reason we didn't do it much. One seining trip I remember was probably my fault. It started one Saturday afternoon in an incident totally unrelated to seining. We were having a family get together, with a hamburger cookout and lots of games like croquet and badminton. Farris and Elma Beasley and their children, Danny, Glen and Bill were there. So were Uncle Allen and Aunt Mary George and Jeannie. Anytime Jeannie was around, I was likely to do something to try to impress her. She was a few years older, beautiful and intelligent. She was also enough of a tom boy to give me all I could handle as I strived to do my best to make her regard me as more than a simple country cousin.

One of my favorite methods for impressing her was my old standard. Catch a snake and do something with it. Jeannie was like Gloria in that she had no fear of snakes, but there was a difference - she was intrigued by them and was willing to encourage me to do silly things with snakes. That day I caught a beautiful green garter snake. It wasn't hard because I knew enough about the wild animals around our house that I always knew where I could find a snake. It was like I knew where the coveys of quail were hiding, and which brush piles were most likely to conceal a rabbit.

My ability to catch snakes had been too long established. It took more than that to impress Jeannie, but she knew how to make me reach new heights of accomplishment. When I showed it to her, she said, "Big deal. It's a garden snake. I've seen a million of those."

I wasn't going to be defeated so easily. I let it wrap around my arm and guided it up my arm. For a moment, I thought she might be impressed, but not yet. I was losing her attention. In desperation I said, "Ever see anybody kiss a snake?"

She simply said, "Bet you won't put it in your mouth."

By now, there were various other siblings and cousins gathered around. Jeannie was the oldest of the this set of cousins, but I was next in line. My credibility as an older kid was at stake. So was my reputation as a snake handler.

Jeannie sensed uncertainty in my hesitation. "What! Are you scared of a little garter snake?"

"Not scared. It may pee."

"I think you're scared. I'll give you a nickel."

That was too much. I was beginning to think that it wouldn't be too bad. I said, "I'll do it for a quarter."

"Do it."

"Give me the money."

"You won't do it."

"Give it to Alan. He'll hold it until I do it."

She handed him the money and I realized I was trapped. I didn't hesitate. I encouraged the snake to coil up in my hand and quickly slipped it into my mouth. The cousins, led by Jeannie, started yelling and running. "He did it! Warren put a snake in his mouth!"

I was smiling as I spit the snake into my hand and slid it in my jeans pocket. It occurred to me that I had salvaged my reputation as a snake handler at the expense of my reputation as a person of good judgement. As if to prove the last point, I forgot about the snake as we proceeded to the house to find new play challenges.

After all the cousins left, I remembered the snake that I had put in my front pants pocket. I put my hand in and it wasn't there. It had escaped. It had been at least a couple of hours since I'd put it in there and practically all that time had been spent playing with the cousins. The most likely place it escaped was somewhere upstairs because that was where we'd been playing.

In my concern and panic over my lost snake I exercised very poor judgement. I told Mama about the snake. She didn't like snakes, was afraid of snakes, and had never been the least impressed by my ability to find, capture and make money out of snakes. In fact, knowing how much I liked to catch snakes, she had specifically and repeatedly warned me to never, under any circumstances, bring a snake into the house. She had repeated that warning, in Daddy's hearing, that very morning. She probably knew I'd be bound to find a snake as I tried to impress Jeannie.

I hadn't exactly 'brought' the snake into the house. It only happened to be in my pocket as I was playing with my cousins. To me, bringing a snake in meant that I'd consciously brought a snake in. 'Unintentionally' bringing a snake in shouldn't count. In Mama's eyes, it counted. She didn't like that there was a snake loose in her house, and she didn't like that I'd been responsible, despite repeated warnings. When Mama didn't like something, she made sure Daddy knew it.

Daddy didn't enjoy punishing us. In fact, I think he hated it, but he knew when he had to do his duty. The dreaded belt came out and I had to take my punishment. He even included Alan in the punishment, simply because he'd known about it (I think Daddy also wanted to make certain that justice was fairly spread around). We never found the snake and I never brought another one in the house.

The next day, when Uncle Edward came by and asked if we wanted to go seining, it was a Sunday. I never expected Daddy to give up a Sunday to go fishing, but he was almost eager to go. It occurred to me later that the reason for going fishing was for more than to catch a mess of fish. He wanted to make it clear, in his own way, that he still loved us, and the snake incident was past as long as we had learned our lesson. We had. There was additional embarrassment. I had on shorts which made it possible for Eddie and Peggy to see the pink welts on the tops of my legs. Eddie laughed as he said, "Looks like Warren got a whippin'." I mumbled some kind of lie and jumped in the water to hide my hiding.

Cleaning suckers

(Jepp Collier method)

If fishing was good, we would have a dozen or so suckers by the time we quit. If we went fishing on the big Cane, and had a good day, we'd catch forty or fifty suckers. Same thing if we seined (which we didn't do much). Whatever number we caught the next step was cleaning the fish.

Papa Jepp was a good fisherman and he was a great fish cleaner. No man could possibly be more demanding when it came to prepare a fish to be fried. He taught us that each part of every fish had to be carefully washed, scaled, washed, scraped, washed, scraped and washed again, just in case. Usually, each fish would be scaled and scraped by one of the kids, then the little carcass would be "gone over" by Papa Jepp again to make certain that it was impeccably clean. He used warm water to loosen the scales and cold water to firm the meat.

The 'scoring' is an important part of making the bony fish edible. After the fish has been super-cleaned, the last step is to take a razor-sharp knife and slice crossways across the side of the fish where the scales had just been removed. Each cut was deep enough to go almost all the way, but not quite, through the flesh. Each score should be about a third of an inch apart so that the pieces can be thoroughly fried all the way through. Sucker meat is fine, light and delicious. If the 'scoring' is done right, the bones disappear during frying.

The fish are put in salted water and left in the refrigerator for several hours. If desired, the fish can be frozen at this stage, or simply allowed to soak up the cold, salty water. The brining allows the fish to soak up the saltwater, which, no doubt, adds to the final taste experience.

At cooking time, the fish is dredged in salted corn meal. Maybe a little flour is mixed with the corn meal to make it stick better. The lard or shortening is put in the iron skillet, heated to medium hot, so it sizzles when water is dripped in, and the battered fish are carefully laid into the skillet. Nothing in the world smells better than suckers frying. Corn bread sticks cooking in the oven add to the smells and the anticipation.

If the suckers are cooked correctly, there is a wonderful light meat surrounded by a crisp, fried corn meal husk. The bones should be mostly gone, so the diner can focus on savoring the delicious fish. If the chef is properly trained (ie., if she can cook worth a flip), the fish gravy byproduct may come dangerously close to eclipsing the flavor of the main dish.

Sucker fish gravy is the perfect food. It starts with the fish-flavored, slightly scalded left-over grease that the fish was fried in. Pour out all but a few tablespoons of the left-over grease and add some flour. Then add some milk. Mix it some. Add some more flour and more milk until you get gravy. Practice it - you will eventually get the hang of it. Remember, salt and pepper are good things.

The novice consumer of sucker is well-warned to beware of the few, remaining tiny, sharp bones which sometimes glom together and can get trapped in your goozle before you know it. They are too small to cause serious harm, but they can be extremely uncomfortable until dislodged by a piece of gravy-soaked corn bread, or, as happened to me once, expelled by a torrent of projectile vomit.

Sounds bad, but the wonderful taste of the fine, light sucker meat is worth the slight risk. In fact, the experienced sucker eater rarely, if ever, has any difficulty.

I think the best meal that I can think of is fresh white sucker, fried in corn meal batter to a deep golden crispness and served with fish / milk gravy and corn bread. Add some mustard greens or poke sallet with boiled eggs, a wilted lettuce salad, some new potatoes and a big glass of sweet tea and you might as well have died and gone to heaven.

More on Jepp

When Papa Jepp got old, he started having strokes. Sometimes he had trouble walking, sometimes falling, and climbing trees to fish became difficult, then impossible. Somehow, he never lost his touch. He could fish when he could hardly walk. He fished from the banks, which usually is not satisfying because you can't see the fish as well or get the best angle for snatching, but he lacked the balancing skills to climb or to brace himself in the tree while fishing. When he could do nothing else, he even fished with worms, but he wasn't as happy as when he could snatch.

One of Papa Jepp's favorite times was when Uncle Jim would visit for a few days with his wife, Betty, and children, Rock and Julie. One Sunday afternoon, after the usual big Sunday dinner, most of us were thinking about going out to sit on the porch for a spell of visiting. Papa Jepp and Granny had the best porch imaginable for a Sunday afternoon sit down. The porch almost completely ran the length of the southern and eastern sides of the house, and, no matter how hot it was, the southeast corner of the porch had a breeze.

As a child, I marveled at the ability of the grown-ups to sit for hour after hour in rocking chairs, folding chairs, and every other kind of chair that could be dragged out onto the porch and talk for hours. They'd talk about everything imaginable, but most of the topics were about what other people were doing, what was going on in town or at church or the weather. Papa Jepp was always deeply involved in the conversation because he was one of the best talkers around. He knew something about everything and everything about some things. He rarely argued. He simply talked and made the talking interesting. Papa Jepp didn't have to wait for the porch to have a conversation. He'd start a conversation with old buddies at the stockyard or with insurance salesmen who dropped by to try to sell a policy. He'd talk to strangers in the courthouse or the person sitting at the next table in a restaurant.

More than once, I heard him say that, given a couple of minutes, he could find almost always find someone that he and a stranger knew in common. More than once, I saw him do it. He'd start by asking a total stranger he'd just met some simple question, like, "where are you from?" That alone would often do the trick, because he knew somebody in almost every town around, and he knew a lot of people in the many little towns in and around Lincoln, Moore, Franklin, Giles, Marshall and Bedford Counties. If he couldn't find someone in common based on geography, he'd start on profession. If the person was involved in any way with farming, he'd find out what kind of farming they were most interested in, and then he'd start thinking of people in that part of the farming business. Since, at one time or another, he'd been involved in dairy cattle, Hereford beef cattle, horses, mules, sheep, wool, corn, tobacco, hogs, hay, and probably some other agricultural enterprises, he knew almost anything there was to be known about agriculture.

When he met someone new, he was relentless in establishing the basis for a new friendship, and I never saw him fail. Usually, in less than a minute, he and the stranger were laughing and chatting about their common friends, kinfolks, or business interests, which often led to the discovery of more shared experiences. The poor stranger, without knowing it, had been drawn into a conversational trap from which he would likely escape only with some difficulty, and he'd have a new friend.

One Sunday, the grown-ups were heading for the porch, when Rock said something about going fishing. It wasn't a strong suggestion, just a passing thought to break the boredom of a Sunday afternoon. Papa Jepp was crippled up by this time, so the family was quietly discouraging his fishing trips on the basis of the fact that old men in their late 70's who'd had strokes should probably not be climbing trees. Everyone in the family knew very, very well that climbing trees was implicit in snatching for suckers. They also knew that Papa Jepp hated to fish from the bank if a good tree was at hand, and that he'd probably try to climb the tree whether he should or not.

The conversation that ensued illustrates family dynamics as well as any I ever saw. Here was an old man, crippled by strokes, speech halting and slurred, but with undimmed determination to squeeze every possible moment of pleasure into the waning years of his life. He was surrounded by family who loved him, who wanted to protect him and who were unable to make him do anything he didn't want to do or stop him from doing anything he wanted to do. When Rock mentioned fishing, Uncle Jim shushed him fast, like fishing was one of the most horrifying suggestions that could possibly be made, but Papa Jepp heard the suggestion and liked it. He said, "I reckon the water's clear in Rocky Hole, Jim, and there's not hardly a cloud in the sky."

Mama knew how much Papa Jepp enjoyed talking on the porch so she thought she could tempt him away from fishing by saying, "Let's go sit on the porch for a little while."

"We can sit later. Rock wants to go fishing."

Everybody, even the boy, knew who really wanted to go fishing.

Uncle Jim said, "We need to head back to Kentucky before long."

It was interesting how old man Jepp, whose loving nature was always in evidence, could still be firm when he needed to. He simply said, “Won’t take long to go fishing.” We went fishing.

Gigging

Gigging is for both frogs and suckers. You gotta be careful with suckers, though, because if you get caught with a game fishlike a bass in your bag, you could be in big trouble with the Game Warden.

Gigging starts with the gig. You can buy gigs or make them yourself. The bought ones are okay for frogs, but fish gigging sometimes calls for a sturdier design. Also, bought gigs always have barbs and these are not as good for fishing because it is often preferred to use the fish gig to pin the fish on the bottom of the creek and pull it out by reaching down into the water and pulling the wriggling fish up with the gig.

Three gigs made by Daddy (William Gill)

The next key ingredient is a good light. We eventually came to use flashlights, but carbide lamps were our favorite for a long time. The carbide lamp was made of brass and had to be about half filled with carbide before we left the house. When we got to the creek, we'd add water and it would start to fizz. Then we'd close it and the gas would start coming from a small nozzle in the middle of a reflector. When the gas was lit, and the flame adjusted, the result was a bright white light.

Daddy had a strap contraption that he used to mount the carbide lamp to his forehead just like miners used. That was perfect for fish gigging because the light followed exactly where you were facing, and it left both hands free for gigging.

Dogwood blooms mean different things to different people. Some people are taken with the beauty of this twisted little tree. Others find religious significance in its cross-like bloom. Sucker fishermen know that when the dogwood blooms the suckers start shoaling. Shoaling is when the fish, females laden with eggs and males eager to fertilize them, fight their way upstream to their breeding grounds. For fishing purposes, this means that fish can be found by the dozens swimming in the rapidly flowing shallows.

Spearing the swiftly swimming fish is not without skill. It may not be as specialized a feat as snatching, but many a gig has been bent in a missed effort. It is easier to gig a still fish in a two- or three-feet deep pool. The fish will usually be momentarily dazed by the bright light. That moment is the time to strike because the fish will soon move, and when it moves it goes quickly and is likely to be spooked.

The best fishing is still gigging on shoal. Especially the big ones - the old fish that know how to make elusive moves as they tear up the shallows in hopes of making it to the hiding places they know how to find by instinct. Many of the big ones are female, fat with sacks of delicious yellow eggs in their abdomens. If they are at the peak of their runs, there may be several suckers hitting the shallow water at once. They make enough noise that they can be heard before they are seen. This is when the gigging is best. The thrill of the chase is compounded by the sheer numbers and size of the prey and the nighttime eeriness of watching flashing fish slither and slide through running water in the dancing reflections of the flashlights.

There aren't many frogs on the creek. For good frogging you must go to the little farm ponds. There is a longer window of opportunity for good frog gigging - practically all spring. Gigging would seem to require a gig, but a twenty-two rifle also works, or even a four-ten shot gun.

If anything, a strong light is more important for frog gigging than fishing. The hypnotic spell that the light puts on the frog is dependent on a strong beam of light. It is very frustrating for an evening of good frog gigging to be thwarted by weak batteries or a flashlight that refuses to cooperate in keeping a strong beam for at least a couple of hours.

Finding a pond loaded with big bull frogs that no one else has hit yet that year can easily yield twelve or fifteen big ones and several more mid-sized frogs. The frog gigger can hold his own light, or one person can hold the light while the other gigs, the advantage of the latter approach being that the flashlight can be held more steadily by a non-gigger. Taking turns holding the light allows everyone a chance to get in on the gigging action, although the uniquely squishy feeling of stabbing a frog into mud until the gig pierces through its green, slimy body is not so thrilling to everyone. There have been people, mostly girls, who have found that holding the light is their preferred role.

Usually, the light holder moves slightly ahead, the light beam slowly searching the froggy ecosystem at the edge of the pond. They like to sit on their haunches beside or upon rocks, in watery holes created by cows stepping off the bank in search of a drink or on the grassy bank. Later in the spring, they are more likely to be further up the bank. If there is moss or cattails, the frogs are likely to try to find a hiding place there.

If the light holder is careful enough, he can stop the beam dead on the frog. Frogs have evolved to carefully blend into the pond-side color scheme and are very difficult for humans to see in the daylight, but the pupils of their eyes are huge so that they can see at night, and that is their downfall. The beam of the flashlight almost always falls first on the highly reflective eyeballs, so it is not difficult to see the frog at night.

If the beam stays in place, the frog will be stunned for quite some time. The gigger simply moves into place and makes his stab. Sometimes the frog will jump before the gigger makes his move, but often, if the light is held still and the gigger acts with dispatch, the frog will be impaled. From there, the creature is delivered into a sack with its fellows until it is later removed, dispatched with a blow to the head and its legs are separated, skinned and put in salt water for a night's cooling until they are fried for the next day's memorable breakfast.

Getting the skin off frog's legs is not easy. Vise-grip pliers are one of the best inventions to come along for this purpose. Papa Jepp, in his never-ending search to make cleaning game more difficult, also insisted that we dress the back. "There are two good bites of meat on every back," he would say as he spent a good five minutes carefully peeling slippery green skin from a frog carcass. All I could think was how much work it took to obtain two little bites of frog meat. I certainly didn't voice that opinion out loud. Not that I was afraid of incurring his wrath. He wasn't that way, but he certainly wasn't above voicing his opinion, and he was consistent about not wasting anything. "Don't kill it unless you plan to eat it," was a typical statement. I don't remember him saying it, but he'd agree that "You don't ever waste a single bite of good meat."

Drought

Quiet pools, too small for fish, form in unpredictable places. Most of the pools have mud bottoms but they may also be gravel or solid limestone. Shallow water over mud provides a canvass for crawdads to leave crisscrossed patterns as they scurry backwards, pulling their over-sized left pincher with them everywhere they go. Water-bug and tadpole nurseries form, but the creatures must quickly grow and leave because the pools may only last for days before they dry as the water level in the creek drops or they become engulfed in rising water following a rain.

Nothing is permanent about a creek. The pools come and go, the banks change shape, gravel bars move, even bank-side trees lose their grip and are washed away in a flood. Even the water itself cannot always to be counted on. Drought is the enemy of the creek, squeezing the feeder brooks and springs, until, one by one, each quits delivering water until, finally there is no more water flowing. The creek will live by the water in its pools for several weeks after the flow stops. If the drought persists, the pools will dry.

Bank trees are symbiotic with the creek. Their leaves shield the creek from the burning sun, yielding coolness to shaded pools. Tree roots hold the bank soil. Trees are home to the squirrels and birds and countless insects that form the personality of the creek environment. All the creek must do is give water. When water is plentiful, this is not a problem, but drought is different. Each tree needs many gallons of water each day. They are living siphons, sucking moisture, draining the creek, drying the mud, squeezing every drop their roots can find.

The timing of a drought can be important. Spring droughts don't last long enough to truly hurt a creek, at least they haven't in my time. A summer drought is hardest on the creek because it can stretch far into the fall without relief, and the leaves of summer trees need copious amounts of water to operate their green-making chemistry. Fall droughts are bad because they hit when plants and animals most need to be building reserves for winter, but one interesting thing about a fall drought is that trees stop using so much water. One year, in a fall drought, the creek had dried almost completely when the leaves yellowed and fell. Drought makes for dull fall color, and the leaves fall quickly. Interestingly, as the leaves fell, the creek came back. There was no rain, but the flow started back. At first, the renewed flow was a mystery until Daddy said, "It's the trees. They quit using water and the flow came back. That flow is exactly the amount of water that the trees were using before they lost their leaves."

*****

As a drought lingers and grows, the crustaceous animals and frogs bury deep in drying mud. Snakes find cool places to wind into tight coils and bide their time. Fish die.

Little fish become meals for leggy Great Blue Herons and other creek predators, but big fish find the deepest holes they can and try to last if possible. A drought is hard on fishing, sometimes making the fishing bad for several years after a drought, but nothing is permanent, even death. Eventually the fish come back. They always come back. So do fishermen.

Hunting the Cane Creek Valley

I'm going to start by admitting that I have never been what you'd call a successful hunter or fisherman. I liked hunting and fishing or at least I did growing up. I don't remember even thinking that there was a decision to be made about either one. It was just part of what you did if you lived where we lived.

It was a rite of passage for Daddy's to teach their children how to hunt and fish, with grandfathers, uncles, cousins and friends all playing a role. I clearly remember Daddy helping me figure out which was my master eye and letting me shoot a shotgun as part of the process. He took me to creek and threw a stick in the water to shoot at so I could see how the pattern of shot spread around the stick (that is, if I hit the stick I was aiming at). He taught us all how to align rifle sites, and how far shotguns and rifles could be accurately shot. Papa Jepp was about as good as Daddy on teaching hunting skills, and possibly a little more patient.

Warren with Browning Sweet 16 shotgun on the Christmas following his 16th birthday

I got my first gun, a Browning sweet sixteen shotgun, for Christmas after my sixteenth birthday. Alan got the same thing on the same schedule and Gloria got a 20-gauge, double barrel when she was 16. I'm not sure why she got a different gun; maybe because Mama and Daddy thought a 20-gauge was better for girls. No matter because she was a far better shot than me.

In fact, everyone was a better shot than me. I was pretty good at a lot of things, but never good at shooting. Alan and Gloria were both better at anything that required good hand-eye coordination. They were both better than me at anything that involved handling a ball. I could hit a still target with a rifle, but hitting a fast dove was tough for me. Not only am I poorly coordinated, but I'm also near-sighted and I didn't know it until I was 16 when I failed the driver's test eye exam. It wasn't discovered in school because I usually got good grades and when I didn't I was told I didn't study hard enough. I often sat near the front of the class when I could. Possibly so I could see better even though I didn't know I was near-sighted.

I also was better at reading. Near-sighted people can't shoot worth a flip, but we can read our butts off. I loved books about great hunters (think Natty Bumpo, my hero and role model), but couldn't shoot straight. So, the following is motivated by a combination of poor outdoorsman skills linked with a love of reading and writing.

Cats

I'll start with cats. Cats are great hunters, so that's a reason for putting cats in the hunting chapter. Cats also hunt quail and baby rabbits, so they compete with people for game. Hunters often kill cats (explained later) but not for food.

People who study such things say that dogs were the first animal to be domesticated. It makes sense. Dogs are predators with a strictly hierarchical pack. Humans are the same, plus humans are smart enough to expand their "pack" to include dogs because dogs provide benefits. Maybe dogs are also smart enough to know it is to their benefit to join the man pack and give good measure.

The same people who claim dogs are the first to be domesticated sometimes debate when cats came to be domesticated or whether they are truly domesticated even yet. The relationship of farmer to cats is certainly different than with dogs. Cats are almost always part of the story-book farm, but it is often as a tolerated aside, a necessary evil for de-mousing the crib or as pets. Cats never worry about giving good measure. They go their own way on their own schedule. Toms live violent, romantic lives of wide-spread mystery, inevitably limping home after several days' absence with mauled ears and bloody fur. Mama cats stay home and raise litter after litter, often with horrendous losses to be accepted as a part of being a long-suffering cat.

Part of the cloud of mystery that surrounds the feline world of the Cane is the population of feral cats that are rarely seen but are ever present. They inhabit abandoned shacks and brush piles and are skilled at avoiding human contact. There is food, usually in the form of birds and various other creatures of meadow and wood. Subtle differences between voles and field mice disappear in a few crunches of feline teeth. Quail eggs are a delight to cats, but human aficionados of quail hunting understand this feline trait and readily kill cats on sight in hopes of protecting at least a few of their favorite birds.

The problem is that it is difficult to tell a feral cat from a house cat. The hunter that walks up on a cat in a tree trying to catch young squirrels at play only sees the cat. He might look to see if there is a collar, but the presence of a collar only answers the question about whether it belongs to someone or not. If the cat is a squirrel hunter, it is also a bird hunter. If the hunter's thoughts go in this direction, it is likely that the cat faces a quick demise.

There may not be any real difference between a house cat and a feral cat. After all, the term 'feral' infers that a domesticated animal has gone back to living in a 'wild' state. The real question is whether a cat is truly domesticated or simply choosing to live in proximity to humans for its own reasons.

A black kitten, about half grown, showed up at Daddy's house. She was a little shy, but approachable. We put out some milk and messed with it some and she warmed to us, especially to my daughter, Greer. We took her home to Murfreesboro. Greer named her Black Kitty. She grew up that fall and winter, mostly living in the garage, but coming into the house every time she got the chance.

One day, in late winter, Kitty was acting funny, making a lot of mewing sounds and nervously walking around, tail high. She'd come in, but wanted to go out shortly after she came in. Finally, she went out on the porch and paced back and forth while 'talking' cat language. What she was saying was that she wanted some wandering Tom cat to drop by for a visit.

It didn't take long. Within minutes a neighborhood Tom, a big, dirty, white and gray cat made his skulking way into our yard. The courtship was brief and violent. There was lots of hissing and cat yells, but he accomplished his role in Black Kitty's life and quickly disappeared.

A few months later, Black Kitty disappeared. We worried a little and looked a lot. It turned out that she'd found a good nesting area. By the time we'd figured it out, she'd delivered three kittens in a dark corner of William's closet.

One of the kittens was gray and white, a lot like its dad. It was the most stand-offish and had quirky ways of moving around. One was a fluffy coated soft gray with a friendly personality. The last was black with distinct white markings.

After the kittens were walking and playing, we decided to take them to the farm. One of Greer's friends went with us. After we got there, we all went our separate ways, hardly thinking about the cats after a moment of letting them out and telling Papa and Mama Gill about them. Some twenty minutes later, Greer noticed that two of the kittens were missing. A search quickly turned up the gray one, but an extended search failed to find the black and white one.

We looked everywhere we thought a cat might hide, and finally settled on a possible scenario that it had gone down a rain run-off drain. No one really bought the scenario, but we couldn't come up with anything better and we had to get on with other things.

Time passed. No kitten showed up. We did think of some other ideas for how it went missing, but we never figured it out.

The following winter, when the leaves had fallen away and the brush and grass thickets disappeared under snow, I was taking a bale of hay over the big hill to feed to the heifers. I saw a cat. It was black and had the same markings as the kitten we'd lost. The only difference was that this cat was missing her tail.

I stopped the tractor after it got a little closer. The cat stayed in one place, almost as if it were as curious as I, but when I got off the tractor the cat fled across the road and disappeared into the briars and honeysuckle vines on Jimmy and Dean Dunivan's side of the road.

Over the next few months we saw the cat repeatedly, and several others with it, but the black and white cat was the only one that seemed comfortable letting us come near. The others would disappear down into an old groundhog den under a brush pile or simply flee into the woods. Sometimes we put food out, but if we ever got too close, the cat simply left.

As near as we could figure it, our kitten had joined a feral band of cats and was living on the land. We can only speculate how she lost her tail. I say 'she' because she is small and feminine looking, but I am not good at sexing cats from a distance. I missed the gender of both the other two kittens with close examination. I thought both kittens were females, and both turned out to be males.

The real question is how the kitten left us so quickly that day on the farm. Maybe a cat in one of the feral colonies simply came down and stole the kitten. Maybe the kitten wandered away and was found by a hunter female from the underbrush cat colony who took her in and saved her life. Maybe a coyote chased the kitten and had caught it by the tail when a tom from the cat colony came along and frightened the coyote into letting the kitten go. Of course, to complete this fantasy, the coyote would realize that a tom cat would make a better meal anyway, so simply let the kitten go so it could eat the Tom. Bottom line: tom cat gave its life to save the kitten (all but the tail).

The truth is, we have no idea how the kitten got away and became a member of the feral colony. We can't even be certain it is the same cat, but it looked a lot like it. We are sure that there is a colony of cats living a wild existence up the hill from the Cane Creek, and there are probably many other such colonies along the Cane.

Black Kitty ruled our happy home in Murfreesboro, and the white and gray kitten became a big, fat cat that often tried to help me in my writing by walking over the laptop. He really liked curling up and sleeping on the keyboard. He also loved to walk along the piano keyboard, but his tunes were not good.

Pets on the farm

Whatever their role, dogs and cats are integral to the picture-book farm. Working dogs come in several forms, depending on the needs and likes of the farm family. Collies are a nice-image dog, but usually didn't fit the real work of the Cane Creek valley farm. English Shepherds, and in later years, Australian Shepherds, with quick intelligence, herding sense and joyful work ethic are a better fit on working stock farms.

An experienced stockman with a good dog can do things with animals that defy the imagination. Sending a dog to bring in the sheep is routine. The good ones can make the sheep load into a trailer in the middle of a pasture.

Papa Gill was a good dog trainer, and Farris Beasley told us this story as proof. Papa Gill, Daddy, Uncle Edward, and Farris had been in the corn field but broke for lunch. Back then lunch was dinner and dinner was supper. Lunch (dinner) was not a salad with a sandwich – it was a full hot meal with meat, bread, and a several choices of hearty vegetables. As they washed up before dinner, Papa Gill realized he only had one of his gloves. Papa Gill went to the door, and said, "Dog – glove," and pointed toward the corn field. The dog lit out to the field like it was shot from a cannon and came back in a minute with the missing glove.

If there's been a few good shepherding dogs up and down the Cane, there's been more than a few good hunting dogs. Hunting the Cane Creek Valley comes in several forms, but most of them involve a dog of some size, shape or talent.

With a long hair coat and an elegant demeanor, the English Pointer is the aristocrat of hunting dogs. Hunters constantly debate which breed is best, and the close hunting English Pointer doesn't usually win many field trials, but it is a good fit for hunting the brushy fields and Cedar thickets of the Cane Creek Valley. Add the fact that the English Pointer, as a rule, is a gentle dog with a medium level of energy and a lot of hunting heart and you understand its popularity.

The short-haired German Pointer is a high energy dog, ranging far afield with a determination to hurl itself with abandon at the task it was bred for. It is a stirring sight to go out on a cold morning and release a pair of these eager hunters and watch them cover acres of brushy bottom fields, sometimes going in full run and other times twisting and turning to best detect a whiff of prey. They read scents like we read road signs. One odor says slow, another says turn but the smell that the causes the dog to freeze into the classic pointing pose is close-up birds. By coming to a complete stop the dog not only tells the hunter that a covey or even a single bird is near, but also causes to birds to stay put instead of exploding into the air.

It is a neat quirk of hunting instinct that makes a bird dog stop the instant it comes within striking distance of a covey of quail. It is certainly instinct and not learned behavior because pups start doing it when they are just a few weeks old. Let a half-grown chicken walk by a litter of roly-poly pups, just learning to run and play, and see their reaction. At least a couple of the pups will spy this interesting beast and their hunting instinct will force them to explore. Just as they get within a foot or two, they will pause and may even freeze in the classic bird dog stance with a front foot raised and tail held still and raised. A young pup won't hold the position for long, but it will be there for a fleeting moment and indicates that the pup has the natural inclination to point.

Daddy used to put a quail wing on a fishing line and cast it in front of puppies as they got a little older. The good 'uns would point. These would get praise and the pups who pointed would be noted as being worth working with in hopes of making a good pointer.

From this start, the pup must learn basic obedience, but these dogs have loads of intelligence and live to please their masters. Add a little patience from the human and the dog will be ready for the field.

Turning a promising pup into a good hunting dog is a matter of exposing the dog to the hunting experience. They absolutely, totally, and completely love to hunt. You do not have to teach any decent dog to love the hunt. What must be taught is when to hunt hard, how to keep up with both the quarry and the hunter, and not leave the hunter too far behind. It is pointless for a dog to have a point on a quail that the hunter can't see because he has lost the dog. Pointers don't make any noise when they are pointing so the hunter must know where they are at all times. This can be tricky in areas where there are acres of rolling land covered with sage grass (broom sedge for those who like correct names) and briars interspersed with Cedar thickets. The dog that is over-exuberant can get several hundred yards ahead of the hunter and come upon a covey. He freezes into a point and the hunter can't see or hear him. He may have to hold the point for minute after minute. The quail will tend to hold for a short time under the point but will likely start to move after the birds realize that this predatory dog is frozen in place. The dog will hold if the birds stay still, and a little longer, but after they've been on the run for a minute or two, the dog will start to trail. If this is done in front of the hunter, it is an important part of the excitement of the hunt. If it occurs while the hunter and dog are lost from one another, it is frustrating for both.

Experience in hunting is best gained by working young dogs with seasoned dogs in front of experienced hunters. The old dog will know not to lose sight of the hunters. The old dog will know how to come to point at just the right moment to keep the birds from running before the flush. Sometimes the birds will run, no matter what. A smart dog, with experience, knows how break point at the right time and circle the birds to keep them within a good shooting range.

The flush is a time of excitement for both hunters and dogs. There must be an evolutionary advantage in the explosive way that quail take wing. First, you can't see the birds - they are almost perfectly camouflaged. The dog tells you they are there, and you might see one or two little heads. The males have a couple of white stripes on their heads, so are sometimes a little visible, but for the most part they are totally blended into the brown-gray brush and overgrowth.

When the quail decide to fly, they all take off at the same millisecond. It's an explosion of activity. Each bird beats the air with its wings, making as much noise as feather-covered bone and muscle can possibly make. When fifteen or twenty birds do it at the same time, the result is a cacophony of sound and fury. Pin feathers fly loose. The air is literally turned into a drum as they fly in as many different directions as there are birds in the covey. Where once there was nothing is now a swarm of feathered activity. No predator can possibly help but be momentarily put off by the explosive flutter that is going on all around.

The first time a dog experiences a flush has got to be one of the most confusing things that could be imagined. Here, in front of his astonished face, the little birds he has been smelling are turned into so many rushing, percussive little helicopters flying around his head. I hate to admit it, but it has never failed to scare the bejesus out of me. That, plus the fact that I have never been a good marksman (can't hit the side of a barn), explains why I usually get fewer birds than anyone.

A good dog will keep his head during the flush and watch which way they go so it can start working singles. The young dog's first idea is to try to jump and snap at one bird, then another. The birds are too fast and will elude capture, then the dog hears its master's voice break through his adrenalized brain ordering it to heel. "Jake! Heel!" How can master possibly want a "heel'" at this moment thinks poor young Jake, but something in Jake's nature finally makes him come to heel because Master seems a little upset with him.

From there, it is time to hunt singles. This really separates the green dogs from their betters. The inexperienced dog flitters hither and yon. The experienced dog works logically with the goal of setting the hunter up for the best shot. The experienced dog knows that many of the kills are done while working singles, probably because only the best hunters can pick up more than one bird out of the covey flush. Many hunters are as stunned by the covey flush as are dogs. It takes many years of experience before hunters can keep totally cool during the flush and calmly pick one, shoot it, pick another, shoot it and maybe even pick a third and shoot it. Experienced hunters working with novices often recommend the use of a single shot gun until the beginner learns to make every shot count during a covey flush.

Papa Jepp enjoyed hunting almost as much as fishing and had many bird hunting tales. One wasn't so flattering to his sportsmanship. It was during the Depression and he'd been hunting all day and hadn't seen a bird. He was tired, the dogs were worn out, and it was getting dark and cold. Suddenly, his old dog pointed. The dog was pointing toward a brush pile that was on the side of a hill and Papa Jepp was down the hill at about eye level with the place where the dog was pointing. He could see the birds and they were in an unusual configuration - they were in a straight line. He had an idea, but it required that he walk around the pile and get exactly the right angle for his shot. Luckily, the dog held the point and the birds didn't move. He reached the perfect location and shot. He killed the entire covey with one shot - twelve birds. He seemed embarrassed a little by the story, but he told it anyway. Twelve birds in one shot is just too good to keep secret.

My favorite quail hunting story happened one day when Alan and I went hunting after school. We were hunting on the Welch farm across the road (my farm now) above the old dry pond. We were hunting singles. The dogs were down the hill and Alan was maybe twenty feet ahead of me. A single rooster flushed. I saw it first and swung my gun toward it, but I saw I couldn't shoot without hitting my brother, so I pulled up. By then, Alan had turned and saw the bird coming at him, fast and low. Remember, Alan has remarkable reflexes. It's why he was such a good baseball player, pool player and every other game player that requires excellent hand-eye coordination. He can even catch flies, one after another, out of the air as they fly by. His high school buddies nicknamed him 'Spider-Al' for his ability to pluck flies in mid-flight.

Alan snapped his hand into the air and plucked the bird from its flight path.

It happened so fast that it took a moment for what he had done to register in our minds. I ran to see up close if he had really caught it. I think Alan was as puzzled as I was about what had happened. He'd simply raised the hand that had done the deed, looked down and there it was. It was a rooster quail. Alan cupped the quail in both hands to keep it from escaping and we looked at the trapped bird. If we were puzzled, the bird was downright perplexed. It had a good escape plan: fly directly at the predator and shock it into immobility. That would have worked with most hunters, but not Alan. If you gamble, sometimes you lose. The bird gambled and lost.

The ideal kill results in a bird that is dead. The dog picks it up and brings it to the shooter who puts it in the game pouch of his hunting jacket. If it is wounded and not killed, the dog sometimes must chase the bird and capture it without additional damage. It is a very bad habit for the dog to chew the bird (think how this goes against a dog's nature to not chew a bird it has in its mouth). The dog is supposed to deliver the wounded bird to the shooter who humanely dispatches it by pinching the head off, or by breaking the neck of the bird at the base of the skull.

We had never been in a situation where we were faced with an undead, unwounded bird. The choices facing us were simple. You kill it or let it go. I think for a moment we both toyed with the idea of letting it go, but there was no way anybody would believe this story without the bird. It had to be done and was. Alan mashed the neck and sent the bird into its death flutter. We finished hunting singles, but I think we had both lost our concentration on the hunt because we couldn't wait to get home and tell Daddy about Alan's incredible bird catch.

A good dog does a bad thing

My least favorite hunting story was about one of my favorite hunting dogs. The dog was a Brittany Spaniel - not a common breed of hunting dog in Tennessee, but there are a few around. This dog was given to me as a pup by a good friend, Ed Simms. Brittanies are a thick-built, long-haired dog that tend to work even closer than English Setters. Ed's father, Bobby Simms, was a respected local lawyer who liked to hunt but didn't hunt much so their bitch was pretty fat and a slow hunter, but she wasn't bad, and we liked to hunt behind her when she was paired with our Setter, Sue.

When Ed's dog whelped, he sold me a male for twenty dollars and I named it Sam (I was reading Lord of the Rings and the pup reminded me of Samwise Gamgee). The dog grew up to be an energetic hunter, although not a great one. It could find birds but was a little slow about learning the finer points of coming to a point at just the right time to avoid flushing the birds. It was easy enough to tell that this dog

Sam, the Brittany Spaniel, is the one pointing, with Warren and show steer in the background.

wasn't going to be a brilliant hunter, but it had energy and enough innate skill to make it a keeper. During his first hunting season, Sam got better every time we took him out and the patient old dog, Sue, did her best to bring him along. We'd spent much of the off season going over the basics, and I was thinking his second season was going to be a good one.

It was spring. I'd been helping Papa Jepp with the Tobacco and we were just sitting down to dinner (lunch) when we saw his sheep running across the hill. Sheep don't run around in a panic during the middle of the day unless they are being chased. This was very unusual because most dog attacks occur during the middle of the night. Only a stupid dog would attack a flock of sheep in daylight right in front of the farmer's house.

Then we saw the attacker. It was Sam. He was having the time of his life. I'd forgotten to put him in his pen with Sue. To tell the truth, I didn't worry about penning him because it never occurred to me that he would do anything except maybe chase a rabbit or two around the yard. But there he was, only a couple of hundred yards up the hill, chasing sheep, nipping their flanks, racing around the flock one way, then another. He was using more energy than I ever imagined he possessed.

After a moment, it dawned on me what was in store. Dogs that chased sheep were killed. That was the rule and there was no exception. I'd heard Papa Jepp repeat that rule many times and he was not the kind of man who changed his mind about things that he knew were right. In this, he was exactly right and every shepherd who ever lived probably agrees.

He said, "I'll get my gun. You gonna do it or do you want me to?"

My panicked brain searched for a way out, but the problem was not complex and there was only one solution. I may have paused for a second, but not for long. I couldn't see any hope for Sam, but I saw no moral lesson for myself in being the one who did the deed. I answered, "I'd just as soon you do it," I said.

"Then go get him out of my sheep and bring him down here." I ran out to stop my stupid dog. When I got within easy hearing, I hollered, "Sam, come." He hated to break away from his fun, but he stopped at my call. Then, probably just to test me, he started back to his chasing. My next call was louder, and I was mad. Why had this dog been so dumb? Why had I put so much time into a dog that was nothing but a sheep chaser?

This time the dog came to me. I grabbed his collar and dragged him to the yard. Papa Jepp said, "Put him in the shed." I shoved Sam into the dark shed of the smokehouse. Sam didn't understand why I was so mad, and he didn't like being shut up at all. He whimpered as I slammed the door in his face.

Papa Jepp said, "I'll take you home now."

Papa Jepp was kind enough to save me having to be around when my dog was killed. He never talked about it again. I assume he shot Sam and threw him in one of the sinkholes where many another sheep killer had disappeared over the years.

I think the hardest part was telling Alan. He had loved Sam as much as I did. I told him about it like it was something that had to be done, and he wondered why we couldn't just promise to keep Sam penned, or just give him back to Ed. I tried to explain it, but my heart wasn't in it, because I'd wondered the same thing. I also knew that there had been no hope of convincing Papa Jepp into saving Sam. I tried to tell Alan that, but I was too young to have the words to explain how the harsh sentence was justified. At that time, we were two youngsters who'd lost our dog. We'd lose others. Loss brings pain. This loss brought an extra measure of pain, but also the seeds of understanding. Pain is a necessary ingredient, along with hope and compassion, in life's struggle toward balance.

There's always another dog.

There was Tom, an English Setter with almost as much energy as a pointer. There was Jake, a pointer that would hunt until his tail and feet were bleeding. Tom and Jake worked together as a team, with Jake hunting out in front, covering the acres at an amazing clip, often out of sight, and Tom hunting closer but always aware of both the hunters and his hunting partner. Tom was more deliberate, and sometimes found birds that Jake missed, but Jake's speed and excellent nose resulted in more finds. If it weren't for Tom knowing to lead us to Jake, many of Jake's finds would have gone unknown. As it was, the typical flush was found by Jake, who would hold the point until Tom led us there. Tom would either back Jake's point or would circle around and work the birds. Either way, we'd get a shootable flush followed by good working on singles.

Sue was another good dog. She was a good hunter and raised several litters of pups that also had some winners. She was a patient dog who put up with a lot of foolishness from teenage hunters. Boys like to kill things. Boys aren't purists about what they kill if they get to pull the trigger, make a big boom, and see how their skills and opportunity come together to allow them to experience the thrill of the hunt. A good hunter knows you only kill birds around a bird dog, but dogs will sometimes point a hidden rabbit. Frankly, they will point almost any hidden animal. If that animal is anything other than a quail, the dog will sometimes point it for a minute, make sure it is not a quail, then shake it off and break point and go back to hunting. Some dogs will let the hunter decide whether the animal they are pointing is worth shooting. That's what Sue did. If she pointed a rabbit, most typically in a brush pile, we would approach it like it was quail and flush it. Being boys, we were too tempted by the potential kill to let the frightened rabbit get away so we'd pick it off if we could, and we usually got it.

Shooting a rabbit in front of a green dog can ruin the dog. It can turn it into a dog that is confused about what its prey is and will be constantly side-tracked to search for rabbits. We shot many rabbits in front of Sue, and she simply shook it off. We didn't shoot rabbits in front of Sue if Daddy was around (he wouldn't like it), but Sue never betrayed us by chasing or trailing rabbits. Frankly, when we shot rabbits around Sue, I sometimes wondered if she didn't look at us with disdain, like she was wondering why she had to put up with this kind of crap, but she was patient enough to know that boys liked to kill and that she'd just do her job and let the boys do their silly things as they wanted.

Rabbits weren't the only non-bird that dogs pointed. Cows and goat does hide their young offspring while they graze. These babies are carefully hidden, buried in deep grass or in brush thickets. It is difficult to believe how well a cow can hide a seventy-pound calf unless you see it done and a goat kid, or kids if there are twins, can be so well hidden that it is possible to step on them without seeing them. Several times, while hunting on the Welch place, one of the dogs would get a tight point on the west hill where a covey of quail could almost always be found. We'd approach assured that we'd soon be in a flurry of birds, but instead, be entertained by a couple of goat kids erupting from the ground. They were instantly at full speed, bolting from cover and running. They'd slow in a moment after they realized they didn't know where Mama was and start bleating. The doe was never far away, so would soon be answering with her own deeper voice. Doe and kid(s) would soon reunite.

The Welch's goat herd had several Nervous Goats. This breed, sometimes called Stiff-Legged Goats, or Fainting Goats, is known for a unique trait: any sudden, loud noise would cause the muscles in the animal's legs and back to stiffen and the animal would fall. Sometimes they literally fell over, but more often their back legs quit working while their front legs kept pulling the goat until the back legs regained their movement. The effect didn't last long before they would shake it off and be running away. Whenever we'd be hunting and come up on a herd of nervous goats we'd shoot in the air or simply clap our hands and watch the goats freeze and fall. Some fun, huh? Country boys have odd ways of being entertained.

Nervous goats were (and are) most common in lower middle Tennessee because they were brought into Marshall County by a wandering farmer in the late 1800's. They are small goats, with a prominent forehead and short face. The story is that the anomalous stiff-legged trait was found long ago in Belgium and kept because it was considered useful to have one goat that would fall during a wolf attack so the rest of the herd could escape.

Nervous goats have received more attention in recent years. They've even picked up some new names, like Myotonic Goats, and 'Tennessee' has been put in front of the 'Stiff-legged' to come up with Tennessee Stiff-Legs. "Nervous Nannies" is a cute alternative. Vanderbilt did some research into the phenomenon, and Dr. Thian Hor Teh, a friend of mine who paid his way through graduate school by gambling at Keeneland and Churchill Downs, went to work at Prairie View A & M in Texas, and did some research showing that the cashmere yield on Nervous Goats was extraordinarily high.

Someone came up with a one to five scale which rates the degree to which goats exhibit nervous tendencies. Shows and exhibits of Nervous Goats ran into trouble when humane groups decided that watching goats stiffen and fall isn't suitable for public entertainment. Despite all that, Nervous Goats are alive and well and common in Tennessee.

A little more on the hiding trait of cows and goats. It is apparently an inherited tendency that allows the mamas to concentrate on grazing while their newborns are safely hidden. Sometimes, if several calves are hidden in an area, one of the cows will stay close while the rest graze some distance away. This is one of those traits that must make you wonder. It makes sense in that it allows the cow to obtain badly needed nutrients, but the calf (or kid), despite being hidden, is relatively unprotected.

I've wondered about this behavior since the summer day that Daddy put me to bush-hogging one day. I was in the long field that the creek ran through and it had grown up in heavy weeds in several places. Daddy had taken the cows out that morning so I could do the bush-hogging without worrying about the cattle, but neither he nor I remembered about the calf hiding trait. I was going slow through a thick patch of briars and iron weeds when I felt the back-wheel roll over something. At first, I thought it was a rock, but a quick glance down and back showed brown and alive. I was sickened to see that it was a calf. My feet automatically went for the brake, but the split second between seeing the calf emerge from under the back wheel, already injured from being run over by a tractor, and when it disappeared under the bush-hog's into the heavy rotating blade, was just not enough time.

For what it's worth, the calf's life ended quickly. Bush-hogs are designed to be able to cut and chop anything from grass to two-inch trees. In many places they are called 'shredders,' and that is exactly what they do. One ad proudly bragged that if your tractor could run over it, our bush-hog will cut it down. What bush-hogs do for farms is good. What they do for calves is not good. I was devastated as the blade repeatedly whomped the soon lifeless calf. I stopped the blade as quickly as I could and quickly backed to free the calf in some vain hope that the injuries were miraculously survivable. They were not. I got to hear the last breath, but there was no question that the calf could survive.

I didn't know what to do, so I did what I usually did when I didn't know what to do. I found Daddy. He and Alan were working in the front lot. I drove the tractor, with the killing bush-hog to them and told Daddy that I'd run over the calf. I wonder now if my pain was showing because his words certainly were the best possible ones he could have given in the circumstances. He said,

"Don't worry. That's a summer calf out of a heifer. It wasn't going to amount to anything, and she was too young to have it." You'd think that I did him a favor to kill his calf. Every calf had value, and he probably hated to lose it. Even a sorry calf brought money, and money was scarce in those days, but he didn't want me to feel bad over something I couldn't help.

"What should I do with it?" I asked.

"Is it dead?"

"Yessir, but I'll make sure."

"Throw it in the sinkhole next to the creek." Not an ecologically sound solution, but it worked. Except for my bad feelings about the whomping sound of a calf under a bush-hog.

Farming the Cane Creek Valley

Farming has changed along the Cane Creek. In the thirties and forties, it was possible to live on a story-book farm. It probably didn't seem like a good story to those living through it. The Great Depression made it challenging, but a young man and a woman could find a piece of land with a tobacco base, borrow just a little money to buy a mule or maybe a small tractor, and three or four cows. Sooner or later, they'd add a couple of sows, a bunch of chickens and a flock of sheep. Growing and canning your own food was essential. Hunting and fishing were not just fun, but important for food. Children were needed because more hands make chores easier. It was hard work, and some didn't make it.

The applied economics of the day required that livestock and crops be selected so that each played a role in keeping the family going. Tobacco's role is one of the easiest to understand: the golden leaf brought cash. It brought more money with less land than any other crop. Tobacco made farm payments. Tobacco put shoes on farm kid's feet. Tobacco grew during dry weather and grew on the well-drained, rocky hillsides of the Cane Creek Valley. Some years were better than others, but every year brought a crop.

Burley was the variety of tobacco grown in the Cane Creek Valley. There are tobacco varieties that grow in many places around the world, but Burley grows well only in Kentucky, North

Carolina and Tennessee, so it is no accident that these states lead the world in its production.

Warren and Alan's 4-H Project tobacco patch

Growing tobacco was a year-round project. In fact, the yield of the next year's crop depended in large measure on fertilizing and liming in the fall according to a soil test. There was also the planting of a good cover crop in the early fall, after harvest but before stripping time. Papa Jepp used Crimson Clover as a cover crop because his sheep did so well on it. I loved it simply because it was so beautiful in the spring when the flowers bloomed. Daddy preferred wheat as a cover crop, likely because he could hay it or graze it. It can fairly be said that it takes fifteen months to grow a good crop of tobacco.

(I hope some day to find out why crimson clover blooms are red but red clover blooms are purple.)

Even the dead of winter had tobacco work. It was winter when decisions were made about where the tobacco would be raised. Would we go back to the same place as last year or break new ground. Most farmers liked to stay in the same place for three or four years, but a field that didn't produce might trigger a change after only a year or two. If you go more than three or four years, chances of disease increased, especially in the years before modern preventative treatments were available. A typical approach was to rotate crops with tobacco for two or three years, followed by a legume like alfalfa to build the soil for the next three or four years, then a crop or two of corn before coming back to tobacco.

How Tobacco was Grown

Much of life in the Cane Creek Valley revolved around growing tobacco. The best fields were given to growing it and no other activity (except church) took precedence over finishing the chores related to tobacco. Hay didn't get cut until tobacco was set. Fish didn't get caught until the tobacco was suckered.

Much has changed in the forty or so years since tobacco was grown in the ways described in the next few pages. Many of the changes were done in the name of saving labor which made it easier for large crop producers to get larger and made it harder for the small growers to compete because their main competitive edge was that they could use family labor costing little or nothing.

There's a curious balance in this progression. Cheap family labor isn't as available now as it was then, at least among the natives of the Cane Creek Valley. Wives work in town, as do the farmers themselves in many cases. Kids have plenty of things to do - good things, educational and church things - and every one of those things has priority over hard work in a tobacco field. If not for the combination of tobacco production technology combined with cheap Hispanic labor, tobacco would have left the Cane Creek Valley earlier, but the end was inevitable – and good.

Tobacco was destined to go away, regardless of the loss of family labor. Tobacco isn't as popular as it once was - something to do with health problems. Links to cancer, heart disease, and facial wrinkling have dimmed tobacco's 'golden leaf' image. Yuppies don't smoke. Many Tobacco growers don't smoke. When I started in Extension, it was important to remember to put out ash trays at farmer meetings. Nowadays, the few smokers who come to meetings know they must sneak outside to smoke.

When Al Gore attacked tobacco at the 1996 Democratic Convention, he lost support of many Tennessee farmers, especially around the Upper Cumberland. They probably didn't mind him being anti-tobacco, but he didn't have to be so vociferous about it. Tobacco has brought many benefits to Tennessee. It lifted millions of people from backwoods poverty. Many of them don't raise tobacco anymore and wouldn't use it on a bet. Papa Jepp didn't use tobacco and his children (Mama, Aunt Mary Neil and Uncle Jim) wouldn't smoke in front of him even though they all smoked. They all knew he thought it was a nasty habit, and he said so many times. Mama didn't smoke much, but Aunt Mary Neil smoked for years and always hid her smoking from Papa Jepp, as did Uncle Jim.

The relationship of Tennesseans with tobacco has always been love/hate, and Al Gore knew it. Then why did he so aggressively attack it? He lost his sister to cancer - that's a valid reason to explain his anti-tobacco feelings, but many people lose family without losing their perspective. If he had kept a little balance in his approach to tobacco, he would have kept the respect of his political base in Tennessee. If he'd kept that base intact, he'd have carried Tennessee in the 2000 election, and he'd have been President instead of George Bush. Lesson: don't stomp old friends if you want the big one.

More on growing tobacco

Tobacco production has changed a lot during the last forty years. I doubt anyone wants to go back to doing it the old ways, but it doesn't hurt a thing to remember how it was once done.

Growing Slips.

Winter was also the time for finding the best site for growing the tobacco slips. 'Slips' are simply young tobacco plants. To make tobacco fit Tennessee's growing season, it is necessary to start the plants in a protected environment. They must be protected from insects that would attack the tender plants, weeds that would rob the baby plants of nutrients and frost that would wipe them out. Making a perfect tobacco bed that yielded hundreds, even thousands, of tough young plants, took a lot of work. First, the bed had to be level and have rich, loamy soil that would support seed germination yet allow the tender plants to be pulled without damage. It also had to be on level ground, and it was preferred that it be somewhere close to the tobacco patch.

The hardest part was burning the bed. Most folks burned beds during February or March. During this time, there were few days when it was dry enough to get a good burn. After a few dry days, the smoke from beds being burned filled the air. It took a hot fire because the goal of the fire was to sterilize several inches of topsoil so that there were no weed seeds or tobacco eating bugs left alive. To build such a fire required a lot of wood, old tires, and anything else that would burn hot. Since it was done in late winter, the weather was still cold enough that the fire itself felt good, but it took a lot of work

Sometime in the fifties or sixties, chemicals replaced burning for sterilizing the beds. Nowadays, the beds don't even exist. Instead, the young plants are grown in styrofoam trays that rest on float beds on enriched water.

Planting the beds was a little tricky. Tobacco seeds are infinitesimally small - you can put hundreds of seed in a teaspoon. They are far too small to spread by picking up seeds with your fingers and dropping them. The solution to this problem is to mix the seed with something else, then spread in a cyclone seeder. Papa Jepp used to make a fifty-fifty blend of ashes from the fireplace with some fine soil. After making sure he'd gotten rid of lumps and stones, he'd add the seed and mix and mix and mix. As usual, he never let a few minutes of extra work prevent him from seeking perfection. After he was satisfied with the mix, he'd spread the seed and have us walk the bed until every square inch had been completely trod.

The next step was to cover the bed with canvas. The border of the bed was lined with logs or old telephone poles or railroad ties, something that would hold the nails on which the tobacco canvas was stretched. The canvas itself was thin enough to let light through, yet thick enough to protect the plants from frost. Rain went through easily. Each canvas was nine to twelve feet wide and they varied in length, depending on how many plants needed protection. The seeds were planted in early April, and the slips were ready to pull by mid-May, after the danger of frost was past.

Plowing the Fields.

Plowing was also done in late winter or early spring. The land would have been planted in some cover crop in the fall. Wheat was probably the most common, but crimson clover was also popular. Farmers with cattle would sometimes get a little extra good out of the cover crop by letting calves graze it. To keep the cows out, they would construct a creep gate that had openings that were large enough for calves to go through, but too small for cows. This would let the calves supplement their mother's milk with the rich cover crop pasture. It also encouraged calves to forage on their own and made them easier to wean.

Sometime in late February or March, it would be time to plow. Papa Jepp preferred to use his team of Ginger and Jack for plowing. Ginger was a Tennessee Walking Horse and Jack was a red mule. He preferred red mules and even kept a jack at stud for several years that was known for its ability to sire red mules. Jack was by that jack, if that makes any sense.

Plowing behind a team is not a casual affair. It is a struggle of man and beast against the packed earth and it requires effort and teamwork. The term 'team' is correct in this context, because both animals and man must work together. When Papa Jepp said 'yay' they knew to turn right, and 'haw' meant left. There was no debate. It just happened. 'Whoa' didn't mean slow, it meant stop, and they stopped when he said it. They followed his orders and he, in return, never over-worked them or failed to provide plenty of feed and forage.

Most people used 'gee' and 'haw' for right and left. As usual, Papa Jepp had his own ways, and 'yay' worked better for him than 'gee.'

The mule-drawn mold board plow was an engineering triumph, designed to slice through soil and turn a furrow with as little difficulty as possible. One animal was on un-plowed ground, and the other walked the furrow that had been turned on the last round. The man followed, keeping steady downward pressure on the plow handles and guiding the point into the soil at the right depth, ever ready to dodge rocks and stumps that could break traces or ruin plow points.

It usually took several days to plow the tobacco fields. The soil had to be the right moisture. If it was too dry, it was hard to break and if it was too wet, it got gummy and wouldn't fall from the plow surface. When the soil was right, it turned smoothly. The same moisture restrictions apply to tractor plowing, but the tractor operator can't have quite the same experience of actually feeling the plow cut the land and seeing it fold below his feet and smell the rich odor of newly turned ground up close.

The fresh clean smell of newly turned soil wasn't the only nasal sensation that met the plowman's nose. He also usually got treated to several doses of mule flatulence every morning. That smell isn't as bad as one might think, but it ain't roses either.

If you are driving along Highway 231 from Belleville to Fayetteville, you cross a bridge at the Wright farm. Just past that bridge, on the right, is a large field that through much of this century has been in tobacco. Papa Jepp used to grow tobacco in that very field and enjoyed talking about how long it used to take to plow that field with mules. He said, "When I'd start plowing that field, I'd get around once before dinner (lunch), and one more round before supper." They use tractors to plow that field these days.

Setting Tobacco.

Sometime in May, after the danger of frost was past, the plants had to be transplanted from the beds to the field. Since plowing, the field had been disked at least a couple of times and was ready for planting. The tobacco bed canvas had been pulled back for several days so the plants would be 'hardened off' in preparation for the shock of transplantation. The bright green plants were grown to some seven or eight inches in height and were crowded together in the bed. They were so close together that it almost seemed like an act of mercy to pull them from their rich, loamy bed and take them to the field where they would have room to spread their leaves and grow.

Tobacco setting time was one of those times where everybody, old or young, male or female, had to hit the fields and work. Setting started with pulling the plants and stacking them in tubs. The plants were delicate enough that care had to be taken in order to avoid more than minimal damage. At the same time, speed was important because so many hundreds of plants had to be transplanted each day until setting was done. Practiced hands moved among the small plants with skill and precision until there were enough to plant several rows. Then the plants were whisked away to the field. Papa Jepp believed that every minute the plants were out of the ground increased their chance of dying.

While the plants were being pulled by part of the family, someone else would be dispatched to the Cane Creek to fill old milk cans and buckets with water. The bed of the pick-up would be packed tightly with any kind of container that would hold water, but milk cans were best because they could be covered to keep the water from sloshing out as the truck was driven across the rough farm roads which might include stretches of creek bed roads or logging roads to get to the tobacco field.

The tobacco rows were kept straight by stretching a string from one side of the field to the other. Papa Jepp was insistent that the distance between rows be precisely measured. He also insisted on the importance of the correct distance between plants. The string that was stretched across the field had knots tied in it at precise intervals - three feet. Someone would walk along with a stout stick, usually Bodock (more on that later), and 'stob' a hole in the freshly disked ground. Each hole would correspond to one of the knots on the string.

Daddy also used a string to keep his tobacco rows straight but wasn't quite as picky about the precise distance between plants. He also used a hoe to make his tobacco setting holes. Daddy's methods were faster, but his crop was larger - if he was going to get it planted, he couldn't afford to be so precise with each plant. Whether the hole was hoed or stobbed, the next step was to pour water in the hole. A bucket was filled with the water from the creek and each hole got a healthy dollop of water. It was tempting to only put a small amount of water into each hole to make the water last longer and keep down the numbers of trips back to get more water. If Papa Jepp thought the pourer wasn't getting the hole wet enough, though, that poor person was in for a lecture about how important it was that each plant have enough moisture to survive. If the plants didn't get a good drink, they might die. If enough plants died, the crop would be small. If the crop was small, we might not make any money. By the time he finished, it was clear that a skimping on water could lead to another Great Depression.

After the hole was watered, we would come along with a basket or small bucket carrying the plants. One at a time we'd gently place the plant in the hole and use our fingers to gather and pack the soil around the plant. 'Don't leave any air pockets,' we'd be warned. 'The roots need to be in wet dirt.'

It could be done by stooping, but young backs soon got tired and we'd be crawling from hole to hole, planting the young tobacco slips. It was a little disheartening to look back at the rows that had been planted only an hour or so before and see that the hot spring sun had caused the tender plants to wilt onto the ground. They looked dead, but we were reassured by the grown-ups that, in a couple of days, most of the little plants would be standing back up. 'Tobacco plants are tough little boogers. They'll make it if they have half a chance.' We had to watch as we set the slips to make sure we didn't plant one with a broken stem. That was one problem that they couldn't overcome, and it happened often enough with young hands handling tender slips.

The advent of tobacco setting machines was a major improvement in our lives. These machines could be pulled by tractors or mules and had a water tank, which eliminated the back-breaking water hauling job. The tobacco setter also had rubber cushioned jaws that open while the seated operator placed a plant in the right place and the jaws gently closed and a chain conveyed the plant into a furrow that had been opened and watered by the machine. After the plant was in place, wheels packed the soil around the plant. Usually, a kid follows the machine with a few plants in hand to fill in occasional misses or to fix the occasional plant that the machine doesn't get exactly right.

The last part of tobacco setting came a few days later. That's when we'd go through the tobacco patch looking for any plants that hadn't made it. Most of the plants would be standing strong, but there were always a few that just weren't strong enough. These would usually be dried green wisps that would be replaced with a new strong slip. Sometimes, a plant wouldn't be completely dead, but just wasn't as healthy as it ought to be. Usually, that one would be heartlessly ripped away to make room for a healthier plant.

It was important to growers in those days to grow as many pounds per acre as possible. This was linked to the tobacco support program. A part of the program was the careful regulation of the amount of tobacco which could be grown. This was an effort, mostly successful, to stabilize the prices farmers received by controlling the supply. Each farmer was allowed a certain acreage on which tobacco could be grown. Inspectors were paid to check the tobacco patches shortly after the tobacco planting was finished. Any amount of tobacco over the allotted acreage was destroyed.

One time, while Papa Jepp was on an errand in Fayetteville, an inspector came out and measured Papa Jepp's tobacco and destroyed quite a large area of over-planting. The inspector was a local high school teacher named Harry Alderdice who probably only did this job to supplement his meager income. He undoubtedly had several fields to do in a short time, but that didn't stop Papa Jepp from being outraged. When he found his dead plants, it was like he'd been personally attacked. He'd known he'd over-planted - he always did - but his strategy was to convince the inspector to destroy plants that were in the poorer producing parts of the field. He might even ask the poor man to pick a few plants from one edge, some more in another location, then ask him to re-measure and see if he couldn't stop before he had to go somewhere else and pull some more. It occurred to me that Mr. Alderdice may have been glad that Papa Jepp wasn't around when he came to measure the tobacco. He may have waited until he was sure Papa Jepp wasn't around and hurried to do his duty before he got back.

It was the maddest I ever saw Papa Jepp. I remember him almost shaking with anger as he told Mama that he was going to find Harry Alderdice and give him a piece of his mind. As Papa Jepp pulled away in his old rattle-trap Dodge truck, Mama was frightened that there would be a fight. Luckily, Mr. Alderdice wasn't home when Papa Jepp got there the first time, and, by the time Papa Jepp finally found the poor man at home, he had cooled down a bit. Even so, Papa Jepp told Harry he didn't appreciate what had been done and that he'd better never step foot on Papa Jepp's farm again.

I suspect that was fine with Mr. Alderdice.

It was standard procedure to destroy all tobacco on the farm that was not in the allotted acreage, including the left-over bedding plants. One year, an inspector forgot to destroy Papa Jepp's bedding plants. After he left, I wondered why Papa Jepp was so happy. He went to the bed and started working it with his hoe. I thought Papa Jepp was simply finishing the job that the inspector had forgotten. What I failed to notice was that he didn't really destroy the plants, but simply thinned them so they'd grow better. I eventually figured out his strategy when I watched him harvest the plants which had grown in the bed. That was the year when he got his picture in the *Elk Valley Times* for being one of top three producers in the county.

Several years later, it occurred to some bureaucrat that the method of controlling tobacco production by limiting acreage was imprecise. The argument was that it would be much more effective to set limits based on poundage. There was to be a tobacco referendum to allow the farmers to vote on either acreage or poundage. Papa Jepp was upset. He'd made it his business to grow as many pounds on each acre as could possibly be done. This was at the root of his anger with Mr. Alderdice and explains why he was so exacting and demanding during the planting and raising of his tobacco.

Many smaller producers agreed with Papa Jepp, but many larger producers, especially the ones in Kentucky and North Carolina, wanted the poundage system so that they could achieve optimal production without having to manage so carefully. They'd simply plant enough acres to make their poundage, and not worry about making each plant count.

Papa Jepp thought the 'poundage' idea was roughly equivalent to communism and would result in loss in tobacco quality and would ruin the industry. He knew it would put him at a serious disadvantage and campaigned vigorously against it. The campaign was bitter, but, in the end, the 'poundage' system won out. No longer was it necessary to squeeze every pound of each acre. Papa Jepp didn't change his farming methods one bit.

*****

A good, slow rain was much appreciated in the days after setting. Papa Jepp said that tobacco didn't need much rain but must be planted in soil with good drainage. Tobacco has a tap root that goes deep to find any available water. He'd say, "I can raise a good tobacco crop from setting to harvest with only three rains." When his audience expressed doubt, he'd add, with the triumphant smile that appeared when he was about to say something clever, "The trick is, I get to pick the timing of the three rains."

Summer Tobacco Jobs.

There was always something that needed doing with the tobacco. After the plants had gotten a good start, so did the weeds. Papa Jepp would give us hoes to work out the weeds close to the plants while he used Jack to pull a cultivator to clean up the middles of the rows. Jack never stepped on a tobacco plant. I can't help but wonder at the whipping he got during his training days when he'd accidentally stepped on a tobacco plant. By the time I came along, I think that Jack would rather have died than step on a plant. It was part of his mule life philosophy.

Hoeing an acre of tobacco is as close to torture as a boy could possibly have to endure. It was hot, dirty work and each row seemed like it was a mile long. We'd start strong, but the quality of our work would wane as the day drug on, at least until Papa Jepp would come along and fuss and make us re-hoe a poorly done row.

God help us if we ever accidently chopped down a plant. I remember doing it once and jamming the plant back in the ground, hoping he wouldn't notice. The trick didn't work. He pointed to the plant at the end of the day and I figured I was in for a lecture, but he wasn't that predictable. He simply grinned, almost as if he'd become part of the game, and said, "I guess cutworms are bad this year!"

If he couldn't think of anything else, Papa Jepp would send us out to look for tobacco worms. Tobacco hornworms are as large as your thumb, have a horn-like protuberance from their rear end, and are about the same color as tobacco. That protective coloration helped them survive predators, including boys. To find them, you had to look carefully at each leaf, top and bottom. When you found one your reward was catching the wiggly, squishy, icky worms with your fingers.

Sometimes the hornworms were covered with small white eggs. Papa Jepp told us that a little wasp laid those eggs and we could leave the hornworm alone if they were toting a load of eggs. He explained that a hornworm with eggs attached was going to die anyway and having more wasps that killed hornworms was nothing but a gift from God.

Even in those days there were poisons that would kill the worms, but chemical control cost money and boy labor was free, plus they needed to have something to do to stay out of trouble. He'd send us out to find the green worms and pull them off. The plan then was to drop them to the ground and stomp them. It was a yucky job for bare feet, and we preferred to throw them down hard enough to kill them. That was acceptable, if the worm was dead. Naturally, I soon discovered that the worm was just as effectively dead if it was thrown against the back of Alan's head. For some reason, Alan didn't like having green worms slammed against his head and he was always a better pitcher than me, so I could count on losing any battles that involved throwing. Losing the battle of tobacco worms wasn't pleasant.

Suckering came next. Suckers are the little shoots that grow from where the leaf meets the stalk. If left in place, the suckers grow into a small replica of their parent plant, but they are considered a problem because they slow the growth of the parent plant and cause problems during curing. The solution in those days was to simply go through the field and pick the suckers out by hand. Since suckering was done after the tobacco plant was some five or six feet tall and the leaves were well developed, the job was hot (little air stirred through all those leaves) and sticky (tobacco leaves are naturally sticky). It was almost as dreadful a job as hoeing. Guess what I discovered: a healthy sucker is a great weapon to throw at your brother. Guess what Alan did: he threw a sucker back at me. Guess what Papa Jepp did: he got mad at us and ordered us to settle down and get to work. A lecture about taking pride in our work followed.

MH-30 is a chemical that stops tobacco from making suckers. If I ever meet the guy who invented MH-30, I'll buy him an ice cream cone. For anyone who is shocked that chemicals are used on tobacco, I suggest you avoid tobacco.

Topping is the last of the summer jobs before cutting. Topping is the removal of the tobacco flowers from the top of the plant. Ideally, you simply walk through the field with a sharp knife and snip the top off. Trouble is that the plants don't all flower at the same time. Some tobacco producers wait until most of the plants have bloomed, even though that means some of them have grown big masses of flowers. Papa Jepp believed that all those flowers were doing was robbing weight from his precious leaves. His solution was to go through the field for five or six or more times to take out the flowers as soon as they peeked out from their leafy hiding place.

Tobacco flowers are a pretty pink and white; they even have decorative varieties called Nicotinia, but they never got big enough to be pretty if Papa Jepp was around. If you wanted to enjoy looking at tobacco flowers, you'd have to go to somebody else's tobacco patch.

Priming is the picking of the tobacco leaves from the bottom up as they ripen. This isn't widely done with burley because it is unbelievably labor intensive and requires serious stooping. Nevertheless, it was something that Papa Jepp thought was a good thing for youngsters to do. His idea was that 'priming' was a good way to introduce kids to the joy of raising tobacco. As usual, he was right. We'd waddle through the tobacco and pick the ripe yellow leaves and pile them in a Western Flyer wagon that we pulled along with us.

Papa Jepp liked for us to do the priming because he realized that the lower leaves would be lost if we didn't do it. The lower leaves simply dried up and ruined if no one picked them because the upper leaves were still ripening. Most people simply let the lower leaves disappear while the rest of the leaves ripened. This approach, which resulted in losing a few cents worth of tobacco, was not acceptable to Papa Jepp.

After we'd pulled a wagon load of ripe leaves, we'd take them to the shade and string them on baling wire. The resulting 'string' of leaves was given to us and we got to sell them as a separately graded lot. It brought fifty or sixty dollars, which was a nice piece of change for a kid in those days.

The most memorable part of the process was the fact that Papa Jepp guided Alan and me to stand beside our basket of primed tobacco so the buyers would know that these farm kids were the proud owners of this lot and were deserving of an extra nickel or so for their efforts. It probably worked.

Cutting tobacco was the hardest, dirtiest, stickiest, hottest and even most dangerous job you can imagine. It was hard because there is a lot of lifting and most of the cutting comes in August and September, inevitably on the hottest days of the years. It is sticky because tobacco is sticky, and never stickier than at cutting time.

The danger comes in many forms. One is tobacco poisoning, which comes from being close to that much nicotine and the other chemicals that tobacco exudes. Housing the tobacco is also dangerous. Many of the barns that people hung tobacco in were old and run down; some even leaned like they were working toward falling – which they were - slowly. The tobacco is hung on wooden timbers placed in parallel rows, which could consist of sturdy logs or could just as likely be rickety old poles. The rickety poles could turn, break, or shift loose. If any of these happen while a someone is standing on them, they are in for either a short fall if he catches onto something or a long fall. The good thing about a long fall is that the person may get out of hanging tobacco for a few weeks, or maybe forever.

The peskiest danger is wasps. Tobacco barns are havens for wasps. Most people try to find the wasp nests and kill them before housing starts, but sometimes a big nest of red wasps or yellow jackets gets looked over. There is nothing at all fun about being attacked by a few dozen angry wasps while you are balancing on an oddly shaped pole forty feet up next to the hot roof of a barn with no good way to escape.

It is smart to carefully inspect a barn for wasp nests before housing any tobacco. Wasp nests come in many shapes and sizes – some only an inch or two across and others as much as a foot wide. There are many kinds of wasps, but the ones that caused us problems were yellow jackets and red wasps. I've heard people discuss which kind is worse, but I think they are both horrible. Some people, like Mama, are allergic to wasp stings and would swell up and must run to the doctor when stung.

In those days, there wasn't any fancy wasp spray in a can to kill wasps, so most everyone had a favorite method. Papa Jepp would wrap a rag around a tobacco stick, soak it in kerosene, light it on fire, and stick the flaming torch to the nest. Another way was to load a sprayer with gasoline and spray the nest which led to a satisfying instant death if you hit 'em right, but if one or more got away you could get a nasty sting. Daddy had a simple way to handle all but the largest wasp nests. He simply reached up and grabbed the nest in his hand and squeezed or mashed it against the wall. I know that's hard to believe, but I saw him do it. I've never done it myself and don't recommend it to anyone, but if you want to try it, help yourself.

*****

Cutting day started with finding the old tobacco knife and making sure it was sharp. There are two types of tobacco knives. One is like a modified hatchet, designed to chop the tobacco stalk. The other is shaped more like a small scythe and is meant to be pulled through the stalk. The result is the same either way; in the hands of an experienced cutter, the stalk is severed at the base with little damage to leaves.

The next step was to load a wagon with tobacco sticks. The sticks are taken to the field and 'dropped' down the tobacco rows. Tobacco sticks are squared, four feet long, and are about 3/4 of an inch thick. The best ones are cut from hickory, but many are made from ash or poplar. The sticks need to be cured because green tobacco sticks will bend and fall from their resting places on the tobacco runs. In those days, the sticks could be bought for two or three cents apiece. Papa Jepp never went to a farm sale without looking to see if there were some tobacco sticks for sale. He'd love it if he could pick up a few bundles of tobacco sticks at a price below two cents apiece.

After the sticks are dropped, somebody comes along and drives each stick into the ground using a heavy, wooden, short-handled mallet. Papa Jepp had a favorite mallet that he kept just for that purpose and no other. It was held together with rusty metal straps and was scarred and worn from years of being used to beat tobacco sticks into the ground, but it was exactly the right weight for the job and fit the hand almost perfectly.

The other necessary tool is the spear. The spear is a cone that fits over the tobacco sticks with a sharpened end that pierces the stalk. The tobacco cutter chops down the leaf-laden stalk and picks it up. Then he forces the base of the just-cut stalk onto the point of the spear and pushes down with both hands. The stalk is pierced by the spear and is pushed down onto the tobacco stick. Four to six stalks of tobacco were put on each stick.

A good tobacco cutter is poetry in motion. 'Whack,' the knife cuts the stalk in one swipe. The stalk is smoothly lifted, and the cutter uses a combination of his own muscle and the weight of the tobacco stalk to force the spear through the stalk and the stalk onto the stick. The process is repeated every few seconds. If the cutter is good, he gets a rhythm going. When he hits his stride, he can cut down a row of tobacco almost as fast as it can be walked. Some cutters prefer to cut two rows at once, but it is a matter of personal preference.

Daddy was a good cutter, probably one of the best. He could cut an acre a day when everything was going well. He knew how to get into a rhythm and had the stamina to go all day. I remember that he used to work for a couple or three hours in a row, then take a little break. He'd pull a little pouch of tobacco from the breast pocket of his bib overalls with one hand and some papers with his other hand. In just a few seconds, he'd shaped the papers into a little trough and poured the tobacco down the center. He'd lick the edge of the paper and do a quick roll and have himself a cigarette. Then he'd reach back into his pocket and come out with a match that was lit off his fingernail or by scraping it on his jeans; then smoothly put onto the end of the cigarette. It took him a minute or two to smoke his creation, but he was usually back to cutting before he'd finished his smoke, the half smoked cigarette hanging from his lip as the knife started to swing again.

The tobacco stalks are left in the field for a day or two to start the curing process and to lose a little moisture to make them easier to handle. Each stick is left in the ground with the newly cut stalks attached by their recent piercing and arrayed in an arc about a third of the way around the stick. The stick, encumbered by the stalks, leaned a little, making the entire assemblage look a little like a tent. A newly cut field of tobacco, observed by moonlight, looks like a tent city.

After a day or two in the sun, the tobacco is ready to be hung. The mules are hitched to a buck-board wagon with slanted sides had been attached. One man stays on the wagon while a couple of others hand the tobacco up. Somebody else guides the mules or tractor through the fields.

Picking up the tobacco from the field is hard enough, but hanging it in the barn is the hard part. Sure, it's dangerous, but the real problem is that it is just plain back-breaking work. There is usually one man on the wagon, handing each loaded stick to the next man who is straddling the lowest run. Then there is another man on the next higher run, and, in taller barns, yet another man on the top run.

The debate among tobacco workers is which job is the most difficult and physically demanding. The guy on the wagon must lift every stalk over his head, so he argues that his is the hardest job because he has to lift every piece of tobacco over his head. The strongest man usually gets the wagon-level job. The next man up also has to handle every stick but gets to quickly hang one stick in three and only has to lift two thirds of the crop to the next guy up. The guy at the top (usually the smallest) only handles one stick in three and doesn't have to lift any over his head, but he has to stay in the hottest part of the barn, next to the roof, and he is the guy that usually gets to find the previously mentioned wasps. There is probably no resolution to the debate of which is the hardest job. Whatever the solution, a group of men who have just filled a barn with a tobacco crop have accomplished a monumental task. They deserve to take a break, and it is a while before anyone must worry about the next tobacco-related task, stripping, which won't occur until the weather has turned cool.

Stripping tobacco is simply the plucking of leaves from the stalk. There are things about raising tobacco that are labor-intensive, and other parts are unpleasant, but one aspect of tobacco production that people can enjoy is stripping. Most of the stripping is done in the fall when the weather is cool, so it is a comfortable job, but the pleasant part is that, while it is busy work, it doesn't stop conversation. In fact, it is very conducive to conversation, hours of conversation. Stripping time is simply another excuse for delving into the never-ending mountain of conversational topics.

Stripping sounds easy but it has its own challenges. First, the tobacco is cured. The ideal curing weather is alternating periods of wet, rainy days and dry days. If it is too dry, the leaves will cure unevenly. If it is too wet, the curing is slowed and may leave spots. Good curing results in an evenly colored leaf prized by buyers.

After curing comes stripping. Before the leaves can be stripped from the stalk, the tobacco has to come 'in case.' This means that the leaves must be moist enough to be handled without shattering, but if it is too moist it may not store as well and could be discounted by buyers.

Wet weather brings tobacco in case, so stripping doesn't usually start until there have been a few days of rain. If the tobacco is ready to be stripped and the rainy spell doesn't happen, it is permissible to wet the tobacco by using a hand sprayer, but the result is not as uniform or desirable.

If the tobacco is in good case, but about to dry out, it is possible to pack the unstripped stalks into piles and cover them with plastic to keep them in good order for stripping.

Stripping usually starts in late October or early November. It usually isn't cold at the beginning of stripping, but there are always a few bitter days before stripping is finished. Ideally, there is a stripping room, or some small room which could be heated, where the stripping operation can be moved when cold weather hit. If not, fingers get cold. Papa Jepp wouldn't use gloves but allowed us to. He preferred the kind of gloves that had the fingers removed.

As the leaves are stripped, they are graded. The bottom leaves, called trash, are taken first. These are the oldest leaves, often ragged and damaged, and are the least valuable. Next comes a couple of grades of high-quality tobacco, called lugs, then the dark red tips.

The stripping room is an assembly line. One person takes the trash and hands the stalk to the next person, who takes off one or two grades of lugs. The next person takes the tips. Ideally, another person is bringing the stalks, removing the graded tobacco, and keeping everybody else stripping.

Modern stripping rooms have similarities to the ones of the fifties, but there is one major difference. These days, tobacco is put into boxes, compressed and tied into bales. For hundreds of years before the advent of bales, the tobacco was tied into "hands."

A "hand" is formed by grasping the tobacco leaves by their stems, being careful to keep the stem-ends even; then squeezing them into a bundle while taking one long, well-formed leaf and wrapping it around and around the stem-ends until the wrapping leaf is used up except for its own stem, which is neatly tucked into the center of the newly formed hand of tobacco.

The hand is then hung on a tobacco stick conveniently sticking from a nearby barn wall. After the stick is filled with hands of tobacco, all the same grade, that stick will be packed and covered so it will stay "in case" until it is time to be taken to market.

Papa Jepp always had something he could talk about during stripping. First, there was the stripping itself, and the importance of getting the tobacco graded exactly right. When some kid would get a few stems out of whack as they were forming a hand, Papa Jepp would lecture about why it paid to make perfect hands. There was usually some iteration of the statement: "If you are going to do it, do it right!"

When tobacco-oriented topics were exhausted, there was baseball. The World Series was always played sometime near stripping time so there was often an intriguing discussion of baseball, old and new. Some name would come up in a baseball context that reminded somebody of someone of the same name in another context, and the conversation would go in a whole different direction. Stripping room conversations could cover a lot of territory in a day.

Marketing was an enjoyable part of the tobacco business, at least if prices were decent. There was always a market for tobacco. Papa Gill and Papa Jepp told us of the Depression markets where tobacco was worth only a few cents a pound, but that was long ago and the government sponsored tobacco support program had stabilized the tobacco market. Papa Jepp was my primary tobacco role model, but Papa Gill did more for farmers by working hard to help get the support program started. Daddy once got to ride with Papa Gill to Knoxville to give a ham to University of Tennessee President, Andy Holt, as an expression of gratitude for all that Andy (who had been an Ag Economist) and the University had done for the program. The main thing Daddy remembered about the trip was that Highway 70 was the longest, curviest road that had ever been or could ever be built. The trip to Knoxville that takes three hours now took twice that then, and they drove up and back in one day. I hope Andy appreciated the ham.

Papa Jepp always wanted to sell in the first sale and was generally right to do so. The first few days of tobacco selling are usually the best, most lively markets. Farmers usually wore their best work clothes to the auction. Their trucks would be laden with the results of months of hard work, and they'd get into line to have their trucks unloaded in the warehouse.

Tobacco warehouses are not fancy, usually a concrete-floored, corrugated tin-covered building that has lots of room to store lot after lot of packaged leaves. After we'd been in line for an hour or two, we got to the unloading dock. Someone would shove a couple of sturdy low-built baskets next to the truck and each grade would be stacked into a basket. When a basket would get four or five feet tall, it would be removed and another started.

When that load was cleared off the truck, it was driven away and another one was started. After several rows of tobacco were on the warehouse floor, a group of men, including one or two state inspectors, would walk by and inspect each lot. They would write the tobacco grade and the government buyer would put a support price. This was important because this was the minimum amount that the basket would bring. Every basket didn't get a support price. The price support program didn't want to own a lot of low-grade tobacco that they couldn't sell later possibly resulting in trouble for the program.

After the tobacco was graded, it would be some time before the sale. Papa Jepp would spend that time going through the warehouse comparing grades and prices to make sure he'd gotten a fair shake. He was as knowledgeable about tobacco grading as the graders, and they knew it. If he came to question a grade, they'd give him a fair hearing and even adjust it if his argument was sound. I think, in truth, they dreaded seeing him coming.

Papa Jepp (center) with whittling stick watching wool being graded.

The sale was the big event. The buyers were led through the rows of tobacco stacks by the auctioneer who quickly did an auction chant at each basket. The auctioneer usually knew within a cent or two of what the tobacco would bring, and he knew which buyers were actively bidding so it didn't take long for a basket to sell. Usually, they hardly stopped walking unless there was a question or comment about a lot. After a nod to indicate the basket was sold, a clerk would scribble the price on a slip of paper and tie it to the basket. It was difficult for my young ears to tell from the auction chant what a basket sold for. It all happens fast, including the bids, and it isn't important that anybody understands outside the little group anyway. There were usually several people in the group following behind the moving tobacco auction, so it was always couple of minutes after the basket sold before we could get to the ticket that had the price on it.

That little number, a scribbled price on a dangling piece of paper, could mean the difference in whether all the labor was worth it or not. These buyers, usually smoking cigarettes, concentrated on buying the best leaf for the lowest possible price, had, in only seconds, determined our tobacco income for the year. It was a tough business that, thankfully, has mostly disappeared.

*****

An advantage of tobacco was that it sold between Thanksgiving and Christmas. It pumped cash into farmer's pockets at the exact time they needed it. Kids needed clothes for the winter and Christmas was the perfect time for them to get 'em. The first day at school after Christmas saw many a kid with new shoes, crisp new jeans, and a shirt that hadn't been mended even once.

The tobacco money had to be carefully husbanded. Only a portion could go to make a big Christmas. Most of it had to go to make year-end payments at the bank and maybe at the grocery and other stores. Finally, some of the money needed to be saved for seed and fertilizer to start the crop year off right.

Dairy cows

Tobacco was good for once-a-year cash, but nothing beat dairy cows for steady income. They were also steady work, twice a day, every day, three hundred and sixty-five days a year. Many farmers started their day with a dark walk to the milk shed where some of the cows were already waiting. Some cows also had to be called in or rounded up and driven to the barn.

When the door to the milking parlor opened, the same cow was always the first one in and they always came in the same order and went to the same place where their morning feed was waiting. Cows are creatures of habit and pecking order. Every farmer can tell you which cow is "boss," and which are at the bottom of the pecking order. He can also tell you which are troublemakers, and which are his favorites. The favorites are often the ones that could fill the bucket the fullest with the fewest problems.

Hand milking was the order of the day in the Cane Creek Valley through the war years and into the fifties. The person doing the milking was usually called a "milker" but the term got modified when milking machines became common. Then the "milker" could be either the person or the machine. A good hand milker (person) could milk a dozen cows in about an hour. These men and women invariably had powerful grips.

The tools of the hand milker were simple: a bucket, a three-legged stool and a rag to wipe the teats. In those days, most cows didn't give but a gallon or two, so a clean two-gallon bucket was enough for each cow. The milker would start by a gentle pat or by gently saying "saw cow" to inform the cow that he was getting ready to start. I'm not sure why daddy said "saw cow" to calm a cow. Probably because his daddy said, "saw cow." And his daddy said, "saw cow." Get the idea. I wonder how many generations of Gill cattlemen have said, "Saw Cow." Daddy also called cows by hollering "sook" very loudly. I suspect the history of "sook" is similar to "saw."

Most of the cows were glad for the relief, but sometimes a cow needed a flank strap to remind her that her business was to give milk and not be shifting around and knocking over the milk bucket.

Some cows were a little mean and liked to kick occasionally. Hock chains were kept handy to discourage such behavior. Hock chains were designed to snugly fit onto the back legs at the hocks (hocks are like human ankles, but higher up the leg).

The other bad habit cows sometimes developed was switching the milker with a wet, manure-caked tail. The cows might pretend they were innocently trying to swat flies, but that excuse didn't work when there weren't any flies in the barn. Cows may be devious and calculating, but there is a limit to their intelligence. Dairymen should be able to out-think a cow. If they can't, they either need to sell the cow or get out of the cattle business.

There are several solutions to the tail swat problem. One is to tie a brick to the cow's tail. If this sounds like a good idea, give it a try. You won't do it but once.

After cleaning the teats, the milker would start by grabbing a teat in each hand and pushing up to initiate the milk let-down reflex. Some milkers would squeeze the first spurt of milk onto the ground to get rid of any dirt or crap in the teat opening. If there was a cat around, it might get the first squirt into its mouth. The rest of the milk went into the bucket. At first, the sound of the milk hitting the bucket was tinny, but, as the milk covered the bottom of the bucket and got deeper, the sound changed to the muffled rip of a stream forcefully ripping into a pool. There was a rhythm to it that lulled the cow into peacefully standing for the time of her milking. As the udder emptied, a man with big hands could milk two teats at once, one in each hand to speed the job along.

After the last few drops were stripped out, the milker would pour the milk into a specially designed funnel with a filter-strainer that removed any specks of dirt that might have gotten into the milk. Straining the milk was often left to the kids while Daddy kept milking. The milk was strained into heavy eight-gallon steel milk cans with metal lids that were pounded with a wooden mallet for a tight seal. The milk cans were put into a specially designed wheelbarrow and taken to the side of the road and a large, strong man driving a truck with an insulated, box-like bed would come by and stop. The large man would jump out and pick up the heavy milk cans and stack them into the back of the truck. Each can was identified by a number that insured the farmer would get credit for his milk and get his milk check each week.

On a cold day, the cows warmed the barn. Milkers often leaned their heads comfortably against the flank of the cow to get extra warmth. There was a special closeness between man and animal in those days that has been lost as the art of milking has evolved into the modern dairy business. Barns have gotten bigger to accommodate larger herds and cleaner to satisfy milk inspectors. These days, milk is never touched by human hands. Instead, it is artificially suckled by pressurized teat cups into glass lines, then to weigh bottles that record exactly how much each cow produces. From the weigh bottle, the milk flows into refrigerated bulk containers. Every day, a huge trailer truck comes and drains the cooler into a refrigerated tank. The milk is weighed and tested, and the farmer gets a check, usually by direct deposit.

To illustrate how the industry has changed, consider what happened in the nineties when an ice storm stopped the electricity flow to the Cane Creek Valley (and everywhere else in Middle Tennessee). Everyone uses electrically powered milkers. Hand milking is practically a lost art. Daddy went next door to Jimmy Dunivan to ask him if he could help him milk his herd out by hand. Jimmy was thankful for the offer, but probably a little amused as he explained that it wouldn't work. First, the milk would have to be thrown away because hands had been involved in collecting the milk. Second, modern cows give too much milk to be hand milked. The third reason was the most interesting of all: cows have been bred over the last fifty years or so to have smaller teats that fit the automatic milking machinery instead of human hands (or calf mouths). Modern cow teats are simply too small to be milked by hand. According to Jimmy, even an accomplished milker would find his hand and arm muscles locking up before he'd milked more than two cows.

Jimmy found a generator and got the electricity back on and used his regular electric milkers on his poor cows with their painfully distended udders.

My first animal research

One memory of milking days is not pretty and makes me ashamed to tell it. I was probably two or three years old and loved to play around the milk barn. We always had cats around the milk barn and cats have kittens. These were fascinating animals, so curious and playful. We were always drawn to play with them and tease them. One afternoon my favorite kitten was playing on the milk cans. The cans had been delivered back to the farm washed clean with the lids loosely fitted so they made an odd playground for a kitten.

I got an idea. What would happen to the kitten if it spent the night inside one of the cans? It never occurred to me that anything bad would happen. I opened a can, carefully placed the struggling kitten inside and put the lid on. Don't ask me why I did it. For some reason, I think I just wondered what would happen. I found out. Kittens don't live after a night in a can. When I opened the can the next day and found the dead kitten, I was devastated for days. What had I done?

Warren hugging his favorite kitten.

Why was I so foolish? What went wrong? I didn't know all the answers then, but I learned a valuable lesson: don't put animal friends at risk.

An old milk can, possibly the site of the kitty disaster (writing on the side of the can says Kraft Cheese, Petersburg Tenn)

Hogs

Hogs were often the next piece of the farm puzzle. It only took a few dollars to buy a couple of gilts and a boar, and that's about all it took in those day to start a hog farm. If you look at a modern hog production operation, you might assume that hogs need to be raised in climate-controlled houses on slotted floors with flush systems and provided with carefully regulated diets. Not so. Hogs are a little like cats from the standpoint of being able to take care of themselves. You give 'em a little acreage and some woods and they'll survive, maybe even thrive. You give 'em a little extra shelter and some corn every day, and they'll raise you some pigs. You give 'em a good worming now and then and a good place to farrow (give birth) pigs, and they'll start giving you enough pigs to help make payments on the farm.

Good hog managers, the ones who found ways to make sows wean off seven or eight pigs apiece and grew those pigs into fat slaughter hogs on cheap, home-grown corn and found ways to sell those fat hogs for good money, were able to make real money. All they had to do was learn to put up with animals that tore down fences and dug up the lots where they were kept by constantly rooting for grubs and worms and whatever else they could find in the dirt.

Some people called hogs mortgage-lifters. When the hog market was high, they sure could help make a payment or two.

Flip side, if the market was down, you could always eat 'em, at least a few of them. Hog farmers on the Cane were not likely to starve.

These days, when people have little direct contact with hogs and much of what they know about hogs is from children's story books or occasional pictures on TV, it might be tempting to think of pigs as cute and friendly. Little pigs are cute and friendly, but grown hogs are another story. Boar hogs (adult males) are large, strong, smart and can be mean. Unless removed at birth, they can have tusks that can grow long and curved and dangerous. Papa Jepp taught me that you had to separate boars from sows with baby pigs because the boars are likely to kill the piglets. Domestication may have made them a little tamer, but they can still reach down to some ancient well of fierceness.

One of the scariest moments in my life was when I looked up the hill from the barn and saw a huge sow (adult female) running down the hill right toward me. She was making a loud whuffing sound and was looking at me like she wanted to kill me. She was running too fast for me to get away so in the most commanding voice I could muster, I said "Soowee." Like "sook" and "saw" with cows, "soowee" is how you talk to hogs. Soowee is usually used to call hogs, but it was all I could think of, so I yelled and stepped toward her waving my arms. Whatever I did, she stopped and stood there looking mean. Later, Papa Jepp told me that she was mad because I got between her and her piglets. Valuable lesson. Don't mess with new Mamas, no matter what species.

Killin' hogs

Hog killing day was special. For one thing, it had to be cold. The colder the better. Usually it was around Thanksgiving, but only if the weather was bitter cold and had the feel of being set in that way for a while. Nothing would ruin hams faster than a warm spell that hit before the meat had taken the salt deep into the center of the ham.

If it was cold enough and the hogs were fat enough, Papa Jepp told Granny to get on the party line and tell the neighbors that he was going to have a hog killing. Some of the neighbors also had a hog or two they could bring, and Papa Jepp would have at least three or four he'd been feeding. Killing and processing hogs is hard work, and having many hands helped make the work go faster and easier.

The Cane Creek played a role in hog killing, because water was a necessary ingredient in the process. Usually, the hog killing was done next to one of the small stream tributaries of the Cane. A fire pit was dug near the stream and a large vat was placed over the fire and filled with water from the stream or a nearby well. Someone had to get up early and build a fire to heat the water to a temperature that was hot enough to scald the newly killed hog so that the hair would be loosened for easy scraping.

Folks would start gathering around the scalding vat a little after daybreak. Some would be rolling cigarettes and others would be trying to figure out the best way to put a bullet in the head of a fat hog that was in the back of an old Chevy pick-up without taking a chance that the bullet would go through and put another hole in the rusty, battered truck bed. Papa Jepp would usually tell 'em to unload the hogs at the barn and put them in with the rest of the hogs. Then there wouldn't be any worry about shooting the truck.

The women gathered to visit in the house and get the knives and pots and pans ready, the lard stands rinsed out and drying, and the sausage grinder cleaned and mounted on the meat table. There were black families and white families, and all wore dress-down clothes because this was a day of dirt and blood and hog fat that found its way on dresses and shirts and britches no matter how careful or persnickety one tried to be. Everyone had jackets and jumpers or some kind of cozy garment, but these were often shed as the work commenced because there was a large measure of warmth in the hard work of the day.

The day was usually overcast, maybe drizzly. If it was raining, the event was put off for a day or two. A trace of snow in the air was okay because it was too early for the big snows that would make hard going. A clear, cold day was the best.

Shootin' the hog

The hogs themselves were crowded into a pen next to the barn. Papa Jepp was the consensus choice for shooting the hog. It was best if the hog was instantly dropped with a bullet to the brain and Papa Jepp was the best marksman. Even more, he'd done it a thousand times and knew the exact spot on the hog's head that was most likely to result in a quick end. He would take the twenty-two long rifle single-shot gun and walk up to the hog, usually trapped behind a heavy gate held by a couple of men. He got close enough for a quick kill, but not right against the head. If he was too close, the hog got more nervous and struggled to get away and the bullet might do too much damage. He aimed at spot just a hair lower than exactly between the eyes.

I saw him kill many a hog and only saw one miss. It wasn't exactly a miss, but the bullet must have gone a little off center and didn't kill the Duroc directly. The fat hog went off in a huffy, bloody panic. There was some discussion about trying to catch it, but Papa Jepp said we'd just let it be and do another hog and take care of the hurt one later. With that, he walked over and put the gun to the head of a big Hamp and dropped it, whump, straight down.

Papa Jepp took a boning knife and quickly cut the animal's throat. He was confident with a knife and cut sure and deep. Red-black blood gushed from half-severed hog's neck. At first, it was a dark flood, but quickly slowed to steady stream, then heavy drips.

If everything was going as planned, there would still be enough of a heartbeat to help pump the blood onto the ground.

While the animal bled, the skin was slit open on the back part of each back leg, exposing the tendons. A stout, pointed stick was forced behind the 'leaders,' giving a convenient hanger by which the hog could be pulled and eventually hoisted onto a pole

Uncle Jim had old Jack, the mule, standing by the gate, harnessed with a singletree ready. When the stick was in place, the chain was attached to the still-bleeding Hampshire hog and Uncle Jim led the mule out. Jack had to lean into the traces to get the heavy load started but was soon pulling the three-hundred pound hog along the frozen ground to the crowd of people waiting to start working on the hog. A trail of blood followed the hog to the scalding box.

I almost forgot the red Duroc hog while we started working on the Hamp, but Papa Jepp didn't forget. When it was time to get another hog, Papa Jepp said, "Look yonder." The Duroc had gone over to a Bodock tree, laid down and was close enough to dead that we were able to easily snare it with the tow chain. Papa Jepp's only miss wasn't a miss after all.

Scalding and Scraping

The water in the scalding box had been heating since daybreak. It was just short of boiling hot. Steam was swirling from the surface. The galvanized scalding box had been removed from storage in the corn crib the day before and carefully inspected for leaks. Except for one day a year, the scalding box served for storing potatoes, or maybe corn.

It took four strong men to dip the hog into the large vat of scalding water. Each man took a leg and heaved the hog until it rested on the metal edge of the vat. They slowly rolled the carcass into the boiling water, being as careful as possible to keep from splashing the almost-boiling water on themselves and their neighbors. Someone used a pole to push the hog around and to test the hair slippage. When the hair started loosening, the pole was used to maneuver the hog into a position where the men could pull it onto a scraping table.

Scraping took only a few minutes if several people helped. Big butcher knives were mostly used, but smaller knives sometimes worked better in the tight places. The hair dropped off in big, wet piles onto the ground. It was the only thing besides blood and gut contents that was thrown away. Papa Jepp made sure we worked fast because he wanted to get the hog hoisted and bled quickly so all the blood would drain. He didn't care if we got the hair off the head and feet because we'd get more time to work on those later.

After the hog was scraped, and Papa Jepp had given it one more good going over to make sure it met his standards, the hog was chained to one end of the singletree, and hoisted up a pole using mule power and chains to do the heavy work.

Someone would step up and finish cutting off the head. Head removal was a combination of cutting by one man and twisting by another. A hog head is some twenty or thirty pounds of good eating, except there is a lot of bone to throw away and a lot gristle to figure out something to do with to get the good out of it. The head would be clumsily flopped back into the scalding vat where it would bob around for a minute or two before it would be grabbed by a hook and pulled out and one or two of the women or children would be assigned the task of scraping the rest of the loosened hair from the jowly hog head.

Papa Jepp insisted that every hair be scraped off, even on the head. He sharpened several old butcher knives just for that purpose. Scraping was made easier by the scalding, which loosened the hair so that it tended to come out with a single swipe of the knife. That theory worked on the body but was less true for the head. There were many little folds and crevices around the eyes and ears and the snout, with its shorter bristle-hair offering its own challenges.

It didn't help that the hog was looking up from slitty, dull eyes accompanied by a slack-jawed, puzzled toothy smile that made one consider how recently this hog was thinking about eating some more corn, or going to snout around the hog lot or come snuffling toward a little boy to show it clearly who was the boss of the barnyard.

Each head took at least thirty minutes to clean, but it seemed like an hour. I became even less enthusiastic about head cleaning after I realized that our family never got the head. The heads went to the black families who used them to make jowl bacon and souse meat. We were usually given a few blocks of souse, heavily spiced with red and black pepper, a few days after the hog killing. Souse takes a lot of work to make, with its countless bits of head meat boiled and boiled until the ligaments turn into gelatin which binds the mixture into a gelatinous loaf when it cools and congeals. Souse not only has a unique taste and texture, but also offers the mystery of trying to figure out what those pieces of meat are. Is that a piece of hog lip or could that possibly be an eyelid?

Sometimes I buy souse at the store, knowing it won't be as good. Bought souse, sometimes called head cheese, has smaller pieces that can never be identified or linked with their previous locale on the hog.

Gutting

The next step was gutting. Care had to be taken to slit the belly without slicing open any of the organs or entrails. Papa Jepp was very particular about cleanliness and the idea of gut contents spilling onto meat was anathema. Even worse would be a burst gall bladder. It was far better to be careful than to be sorry.

He'd start at the top, between the back legs, and carefully slit the belly down the middle. After an initial slit, he'd reach in and find the place where the hindgut attached and tie it off before cutting it. No chance of any hog crap escaping that way. He did the same with the bladder.

Then, as he made the slit longer, the guts spilled into the tub below. He kept the guts as clean as possible, but we'd never eat them. Papa Jepp didn't have a bit of use for chitlings and he considered the heads too much work for what you got out of them. We would usually get back some souse meat from the heads, but the intestines were simply whisked away by the black families, never to be seen again.

The liver was another story. A three-hundred-pound hog has a huge liver and it was one of the parts that couldn't be preserved by salt. In the last years of hog killing, when there began to be freezers, livers could be kept, but before home freezing was possible, livers had to be eaten quickly. That's why the traditional hog-killing supper was fresh liver.

Also, the traditional breakfast the day after hog killing was brains and eggs. Hogs are smart animals and they have a fair-sized brain, although it is buried under a very thick skull. The black families got most of the brains along with the heads, but Papa Jepp would keep at least enough for us to have one meal built around this delicacy.

The flavor of brains is rich, a hint metallic, and very penetrating. If you've ever had brains, you'll remember the flavor (and texture) forever - if you've never had 'em, you'll never understand the experience by reading. Even if you like hog brains, they are a little too intense an experience to take straight, hence the blend of brains with eggs (the ratio heavily in favor of eggs). Maybe it also has something to do with complementary consistency. It is an excellent marriage of foods that each take well to each other.

Children and the occasional adult were sometimes reluctant to eat brains and eggs, and someone always had to say that eating brains would make you smart. Then someone else would say they never did much for the old hog who had 'em before. Granny usually had to make two bowls of scrambled eggs, one with brains and one without. The one with brains had the unmistakable black - gray streaks so there was no chance of a switch, and the brainy eggs disappeared most rapidly when served. Fried tenderloin was usually served along with biscuits and gravy. Tenderloin was another part of the hog that was usually eaten soon because it didn't taste as good after being salted down. We needed a hearty meal because we usually had another long, cold day's work ahead of us to finish processing the hog meat.

Cutting up the meat

Back to processing the carcass. After the guts were taken out, a clean bucket of water was fetched and the inside of the carcass was cleaned. The kidney fat was taken out as part of the cleaning process. After Papa Jepp was satisfied that it was clean, he used an axe and saw to split the carcass down the back.

Each side was taken down and put on a cutting table. The tenderloin was stripped from its snug hiding place, and the slabs of bacon were cut out using both knives and saws. Next, the hams and the shoulders were separated and taken to the smokehouse where they would start getting their salt. Rubbing the hams with salt was another place where Papa Jepp had very particular ways about how it should be done. He didn't think there was any way that you could find a shortcut for rubbing salt into hams. Remember, it must be icy cold or you wouldn't be killing hogs in the first place, and, according to him, the only way to get salt into the ham was to rub it in with bare hands. It wasn't a casual little rubdown either. Not by a long shot. Instead it was a wrestling match between my young hands and the fresh ham. I put all my strength into forcing the salt into the ham meat. Tired fingers jam salt into every crevice and seam. The salt was coarse which made rubbing it against the ham, especially the skin side, feel like a form of sandpaper. Combine the roughness with the cold and the result was young hands turned red and raw.

It was Papa Jepp's opinion that no child ever died from rubbing salt into hams. He was very proud of his record of curing hundreds of hams with only a few ruined, and he credited this to thorough rubbing of salt into the hams before they were laid in the salt box and buried to finish taking up the salt.

Papa Jepp's salt box was a wooden box but Daddy's salt box was actually a large hollowed-out log. The clear advantage of the hollow log approach is the absence of nails. Salt deteriorates nails, making the life a nailed box somewhat limited.

All during the process of cutting up the carcass, pieces of fat were trimmed and tossed into a bucket that would be rendered into lard. Fat was a valued commodity in the days when men and women worked long, hard days and needed the energy. Lard, rendered from hog fat, was the best source of fat because it could be readily obtained, stored for long time, and it gave a unique and wonderful flavor to food. The advent and wide acceptance of shortening was alleged to be better for our health, but it robbed several generations of the joy of eating food with the fulfilling rich flavor of lard.

Think about pies made with lard. The crusts are tender and flaky and complement their filling no matter what that filling is. If you've never had a lard-based pie crust, you only think you've had good pie. Pies made without lard lack balance because the filling is invariably better than the crust. The crust of a lardless pie becomes simply a carrier of filling, a necessary implement for holding the pie together, instead of an integral partner in the taste experience. If you sometimes wonder why pies don't taste as good as your grandmother made, it is because they don't.

Same with fried chicken, except with fried chicken the equation is complicated by the fact that today's chickens are only about seven weeks old when they make the trip to town (chicken processing plant). Today we have flavorless, rapidly grown baby chickens fried in soybean or corn oil. Before, there were chickens of various ages but always at least several months old, with flavor developed over that time by a diet of scratch feed supplemented by barnyard bugs and anything else they could find that could be seen and pecked and swallowed. Take that older, more flavorful chicken and fry it in lard. Fry it until crisp, include a goodly hit of salt and pepper, and you have a taste experience 'that can't be beat.'

As lard has disappeared, so have the lard hogs. The modern hog is prized for yielding lean meat. Rightly so. Today's consumer doesn't want lard and they do want meat that is lean and loaded with red meat. As people have gotten more educated (indoctrinated?) about the threat of fat to our health, it has become more and more difficult to hide fat in other foods. Lean sausage sells, fat sausage doesn't, even on sale. Ninety-nine percent lean slices of ham sell better than country ham.

Papa Gill used to say that "A hog ain't worth killing if it don't give a stand of lard." His favorite breed of hog was Berkshire. Berks were black and white, but their most distinguishing characteristic was (and is) a short, upturned snout. That characteristic made very cute pigs but made hogs that had a short-faced look. Coupling that look with their highly valued propensity to put down fat, and you have a thick animal with a short face. They were also known for having fewer pigs per litter so lost popularity in favor of leaner breeds that had larger litters. This breed and some others are making something of a comeback as heritage breeds that some are raising as the "eat local" trend has caught on.

Poland China hogs also used to be good at getting fat, but they have managed to make the change into lean-type hogs. Hampshires have also made the conversion. Durocs weren't as fat to begin with, so it was easier for them. Same with Yorkshires and Landrace.

Making lard starts with feeding. Hogs fattened in the woods, with diets often rich in acorns, sometimes could get pretty fat, but it tended to be soft fat and the lard didn't harden off as well. Corn-fed hogs made the best lard, so a lot of corn was fed to hogs.

Rendering lard was simply a matter of putting hog fat into an iron kettle and cooking it until anything but lard has evaporated or been skimmed off or settled to the bottom. Cracklin's were the stuff left over. Some people use the cracklin's in bread, but we mostly threw them away.

Sausage

Grinding sausage was easy. Papa Jepp simply took the sausage meat to the store in town and let them grind it in their electric grinder. It was much simpler than using the old hand-cranked sausage grinder. I remember it being used at least a couple of times, maybe for grinding cranberries, but sausage grinding was best delegated to the store. Granny made up her seasoning recipe with hot pepper, sage, black pepper and salt to add to the sausage; then Papa Jepp gave careful instructions to Mr. Gilbert at Gilbert's Grocery about how the job was to be done. It had to be double ground to make sure the seasonings got evenly mixed. Above all, the grinder had to cleaned before his meat went through. As always with Papa Jepp, cleanliness was an absolute requirement.

The sausage came back neatly packaged in cotton sacks. The greasy sacks were hung with baling wire in neat lines from the ceiling joists in the smoke house. Side meat pieces, large rectangles of unsliced bacon, were hung in a similar manner, except they'd had a hole dug through one corner and had been strung with cotton thread.

Sheep

Practically every farm along the Little Cane had a flock of sheep, at least twenty-five or thirty ewes and many had seventy or eighty or even more. Sheep were a good 'fit' because lambing time, when they required the most attention, was in Winter when little else (except school) was happening. Tobacco had been stripped and sold so the tobacco barn was available for nursery duty.

The rams were turned in with the ewes in late summer or early fall as the days got shorter and the weather cooler. This meant that lambing would be mostly in January. Unlike cows, it was common and desirable for ewes to have twins and occasionally triplets. Good shepherds, who bred for twins and kept careful watch on the ewes at lambing time, could achieve one hundred and fifty percent lamb crops, or even higher. This "careful watch" thing is easier said than done, because the most critical watching was during cold winter nights. Often in the middle of the night. Tobacco barns were easily converted into lamb nurseries but were not typically heated. The farmer / shepherd would try to bring the ewes in a day or two before they lambed, but they didn't always show the symptoms of lambing until it was too late and the lambs would be born outside.

I remember many times that Daddy would bring in a lamb or two that had been born outside on freezing cold nights, and had succumbed to the cold before the ewe could clean them off and let them fill their bellies with rich, warming first milk (colostrum). Daddy would check the ewes late into the night and find the almost dead lambs. He'd tuck them, icky sticky as they were, into his coat next to his body and bring them to the house. He'd ask mama to find an old quilt or something he could spread in front of the gas heater and put the lambs down on the warm surface, He taught me how to dry and rub them with an old towel as they warmed near the fire. To me, it seemed like a miracle when almost every lamb would soon open their eyes and start struggling to get up. After he was sure they would make it, he returned them to the ewe as quickly as possible so she could do her bonding thing with her new lamb(s).

By lambing in January, the ewes had all winter to nurse the lambs, so they would be ready for a growth spurt when spring grass and clover was abundant. Most farmers gave the lambs a little extra grain in creep feeders, so they'd weigh in at eighty to hundred pounds apiece by May or early June. Lambs represented most of the income from the sheep flock, but wool was also significant. One way to look at it was that wool covered the cash costs of the sheep flock and lambs represented the yearly profit.

From the standpoint of the farmer who had bills to pay, lambs and wool provided reliable income in the Spring. Selling lambs in May has gotten many a farmer through a cash strapped period.

Shearing

Shearing day started early. The shearers would show up an hour or so after daybreak, and they didn't want to mess around. That meant they wanted the ewes penned when they got there, and they preferred them to be dry and not have their bellies too full. (Sheep with full rumens are more difficult to maneuver – they slosh.) We usually got the sheep trapped in the barn with only a little hay to eat during the night.

Shearing was done in the late Spring. If you sheared too early, a cold snap might leave the sheep without their wool to face the weather almost naked. But the main reason for waiting was to let the "yolk" rise into the wool. Yolk is the word for the greasy stuff, mostly lanolin, that moved into the wool as the weather warmed. It is called yolk because of its yellow color and not only adds weight and value to the wool but also makes the sheep easier to shear. Its like a built-in lubricant that keeps the shears running smooth and keeps them sharp longer.

The shearers showed up in rattle-trap pickup trucks loaded with shears, huge burlap wool bags and tools for fixing up their shearing area. Some shearers liked to work off their trucks, but most would rig an area in the barn with shaft shears and look to the farmer to keep the ewes coming so they could peel the wool off.

The shears in those days were clackity instruments with precision blades. The blades, then and now, consisted of a stationary ten-tooth comb that slides along the surface of the skin and through the wool while the three-tooth cutter slides back and forth along the top of the comb, slicing wool off cleanly as long as the cutting edges are sharp and the shearer knows what he is doing.

In the earliest days of the sheep business along the Cane, sheep were sheared with large scissor-like shears powered by hand muscles. In many parts of the world, this is still the instrument of choice, but not in America and not along the Cane. The first shaft shears were still hand powered, but the power was channeled from a large gear wheel with a crank which some young, strong person would turn. The turning of the wheel drove a shaft which was protected by a metal tube. The shaft was linked to a hand-piece that was skillfully crafted to fit the shearer's hand and allow him to shear the wool from a sheep. As the wool was cut it sprang from the sheep's body is looked like peeling the skin from an orange.

By the fifties, hand power had been replaced by gas generators. They were loud but reliable and effective. As time went on, electric motors replaced the gas generators which quieted the shearing barns to an almost bearable noise level.

The last major innovation was the advent of electric shears, with the advantage that the shaft and the wall-mounted engine were eliminated. The motor of the electrical shears is inside the hand-piece, which makes it a little heavier, but the net result is a machine that is easier for beginners to master and more affordable for people who aren't professional shearers. Professionals still use shaft-drive shearers because they are more comfortable to operate and remove the wool more efficiently. They take considerable training and experience to operate.

Like skinning a cat, there's more than one way to shear a sheep. Papa Jepp's method, the old-fashioned way, needed two people. They would both lift the sheep onto a table or the tailgate of a truck. A rope was looped around the rear legs and someone (often me) would hold the sheep by its front legs and the shearer would direct the holder to turn the sheep as he sheared off the wool. If this sounds easy, I assure you it is not.

One time Daddy had a few ewes that needed shearing. For some reason this group had missed the big shearing day described below and he decided to shear them himself. Alan and I were big boys by then, so he told us to get everything set up so he could shear them when he got back from work. By then I'd helping Papa Jepp shear enough that I was confident I could do it myself so I asked Daddy if Alan and I could shear the sheep. I don't remember if he was skeptical or not, but he gave us the go-ahead, so we went out the next morning and got ready to shear. We had Papa Jepp's old electric shears with a two-inch comb. Later versions all had three-inch combs with are much faster. We finally wrestled a ewe up onto the table and tied her feet. Alan held and I sheared.

This was some level of disaster. We were big boys, but still in our early teens, and weren't really ready for this. A ewe may seem like a docile animal, and actually is normally calm with experienced shearers, but they can be strong adversaries in certain situations, like what we were doing. The first ewe didn't hold still the way she should and neither of us were strong enough nor skilled at positioning the ewe to hold her still for shearing. When we finally got her calm enough, I started the shears and that made her struggle. Alan finally got her settled again and I started shearing. It was only seconds before she moved, and I cut her. Even good shearers make occasional cuts and usually don't worry about it. "Sheep head quick" was the typical comment. But, in this case, I made several cuts before I finally got the feel for it. Even then, I made more than normal. Also, the wool didn't simply peel off like it did when Papa Jepp did it. Instead, it came of in pieces and looked like I was hacking it off with dull scissors. Also, the ewe kept struggling and Alan wasn't holding up very well. He was probably did as well as he could, but we were both out of our league.

That didn't stop us though. We finally got the first one done and I decided the next one would be better. It wasn't. It was worse. By the time we had tortured the wool from the second sheep, Alan was ready to give up. When you're holder quits, the show is over. I said, "Maybe we'll let Daddy help us later."

The other method is called the Australian method. In those days, some shearers used the old method, but, as time went on, more shearers learned the new, Australian way of shearing, which is easier on both the sheep and the shearing. I'll describe that more in the following part.

When the shearers showed up, they did not mess around. Likely they had additional flocks lined up to shear before the day was out. The first thing they did was look around the barn so they could figure the best place to set up their shearing equipment. Most times, on our farm, they used the livestock weighing scales because there was a wooden floor to work on and a tall, wooden side panel where they could mount the shearing motors. After picking the location for the shearing, the shearers almost instantly went to work. First, they'd mount their shear motor and pull out spare combs and cutters. They might do some sharpening and adjusting while one of them picked out a place and put up a wool bagging stand. While the shearers went about their preparations, Daddy would be moving the ewes into pens in the barn so he could move the sheep toward the shearers.

"How many you got, Mr. Gill?" A shearer would ask.

"Maybe eighty-five or ninety."

"I seen a couple of black ones. Want to separate the black wool?"

"Yeah, it'd be best."

"You got any feed sacks?" He was referring to burlap bags that held a hundred pounds of corn or other feed.

"Yep."

"We'll put the black wool in those if you want."

"I'll get a couple and shake 'em out."

"That'll do fine. You may want to turn 'em inside out."

As Daddy did that, the man would turn to a one of the workers and say, "Bring "em on!" Someone would go into the sheep, crowded by a gate into a corner and grab one. If the catcher knew what he was doing, he'd grab it with one hand under the jaw and propel it along with the other hand under the closely docked tail. People who didn't know how to handle sheep would grab them by the wool or try to wrestle them by grabbing their powerful necks. I'll never forget Papa Jepp's advice on this: "It'll wear you out to try to be stronger than a sheep. Use leverage. Guide them with your hand under their jaw and you can do anything with them."

It was the same thing with putting the sheep on its butt. Inexperienced handlers usually try to pick them up and do a body slam. A good sheep handler simply puts one hand under the jaw and the other on the sheep's hip. Simultaneously bending the sheep's head back toward its body while pushing back and down on the rump will cause the sheep to quietly roll onto its tail.

All these sheep handling techniques sound easy. Experienced shepherds make them look easy but the first few times you try them you will find that they are more difficult than the pros make them look. They take much practice and seem to get easier as you grow up and get longer, stronger arms and more leverage.

Shearers were mostly young and muscular, and their backs appeared immensely strong as they bent over the sheep. The muscles in their powerful forearms flexed as they skillfully guided the hand-pieces through the wool.

Most shearers use the Australian method of shearing. This is a carefully developed 'recipe' for controlling the sheep with the feet and legs in positions which keeps struggling to a minimum while exposing the fleece to the shears in a carefully choreographed way for removal. A good shearer can remove an entire fleece in one piece with minimum "second cuts" (small, useless pieces of wool resulting from having to cut an area of fleece twice to remove it).

There is a precise angle at which the comb needs to slide along the skin for correct wool removal and the shears should never be out of wool (with its rich lanolin) for more than seconds to keep the cutters and combs sharp so they don't have to be changed more than a few times a day.

When the shearers hit their stride, the wool almost seemed to roll off the sheep. It looked off-white, even a little dirty, on the outside, but next to the skin the wool was a rich, oily yellow. The lanolin almost dripped from the wool, coating the cutters and combs, keeping them oiled and sharp.

The sheep struggled to avoid being captured, but could not elude Daddy's sure grip, and, when they were settled between the shearer's legs, they calmed. Under the confident skill of the shearer's touch, the ewes became almost complacent, as if hypnotized by the procedure. Each ewe was sheared in only minutes, and was released, naked and changed, to rush back to their ewe-buddies.

My earliest memory of shearing day is one of the most frightening moments of my life. I was very young, not far past the toddler stage. Mama had taken me to the barn to watch the shearing, and I loved it. The huge, tightly packed bags of wool piled next to the barn made for excellent climbing. The lambs, separated from their dams for the first time, were bleating and frightened, but were nevertheless fascinating to me and too small to be threatening as I tried to catch one, then another, always failing.

I had wandered near the gate to the pasture when the shearing was finished. It was time for the sheep to be released, and the gate was opened, but the sheep hadn't been driven to the opening and I had no idea what was happening. All I knew was that there was an open gate and I wanted to walk through it. I was near the middle of the gate opening when the flock saw their escape route. I was the only thing separating them from freedom and a little kid was not enough of an obstacle to deter them. They were going through the gate, whether I was there or not.

Daddy yelled for me to get out of the way. Mama saw what was happening and called my name in panic, but nobody was close enough to help me. The first few ewes ran by me on both sides, in effect trapping me from escaping to safety.

I don't know if I dropped to the ground out of fear or whether I was knocked over by a frightened ewe. Either way, I found myself on the ground, rolled up in a tight ball. Daddy was wading through sheep to get to me but there was no way he could reach me before most of the flock had gone out the gate.

I didn't know it then, but I had two things going for me. One is that sheep have springs for legs and they readily jump over objects in their path. The other is a behavioral principle that if one sheep does something, the rest are likely to do the same thing. The first sheep that saw me on the ground took to the air to avoid tripping over me. The rest of the flock did the same thing. When I noticed I wasn't being trampled, I peeked through my arms and saw a blur of sharp hooves and sheep bellies leaping over me.

I remember hearing Daddy say, "Stay down, Warren"' In only a few moments the entire flock had jumped over me. Daddy had been trying to wade through the sheep, but only reached me as the last sheep cleared me with several feet to spare. He picked me up in his arms and gave me a once over. Not a scratch. I was too stunned to cry, at least at first.

As Mama took me from Daddy, one of the shearers said, "Thought that young'un was a goner."

"He ain't hurt a bit," said another. "I ain't never seen nothin' like that before. How did that happen?"

One of the others, an older man said, "Sheep ain't got enough sense to try to figure out. Let's load our stuff git to our next job."

Horses

A kid being raised on a farm in the Cane Creek valley will learn how to ride. They may learn astride a mule coming in from the fields or when Daddy puts them on the saddle in front of him as he rides after cows. I never heard of anybody ever getting lessons, but, somehow or other, they learn how to ride. I remember getting riding tips from Daddy, Papa Gill and Papa Jepp, and a little later, from Uncle Allen, but I think I may have learned almost as much from the horses themselves.

If I list the horses that most influenced my early years, I'd have to say that Hal, Ginger, and Pluto were the horses that taught me how to ride. Hal was a gelding, by the great Gibson's Tom Hal, one of the foundation horses of the breed. There was a point in the development of the Tennessee Walking Horse when the debate centered on whether the Allen line or the Hal-bred horses would dominate. Allens were generally smaller and showier, with finely chiseled features and plenty of spirit. Hals were bigger and stronger

Pluto. Daddy bought Pluto for $25 and a saddle for $5. Like many Shetlands, Pluto was tough, stubborn and a little mean – the perfect pony for a kid to learn about riding.

with an easy ambling gait that they could sustain all day under saddle or pulling a load. Papa Gill bet on the Hals. The showier Allen line of horses won out over time.

The old gelding that I rode as a kid was typical of the Hal line. He was smart and gentle. He'd come to your whistle and seemed to enjoy a ride as much as the rider. He was great for rounding up cattle and for lazy afternoon rides. He was so thoroughly trained that he was the perfect horse for a kid to learn about riding. I didn't know it or understand it while it was happening, but Hal was probably as much a teacher as a horse.

Papa Gill had trained Hal and could make him hit the best lick of anybody. That wasn't unusual. Warren Gill, Sr., was a master horseman who could make almost any horse or mule do whatever he wanted. He was also an excellent judge and breeder. He kept a stallion or two most of his life and even kept a gaited (Saddlebred) stallion for several years. Daddy said that Papa Gill was the only one who could get that stallion to do the five distinct gaits that Saddlebreds can perform.

One day, Hal was at our house and Papa Gill needed him, so he called and asked me to ride him over. I was home by myself, so I didn't have anybody around to ask permission and, if Papa Gill told me to do something, that was good enough for me. It was almost three miles, mostly along the Chestnut Ridge Road. It wasn't a big deal but I was thrilled when Papa Gill called. Hal was probably about twenty years old, but in good shape and plenty wise to the dangers of traffic.

Hal's only problem was that he was blind in one eye, the result of an accident when someone was driving Hal to a cart and he somehow managed to bang his head into a telephone pole hard enough to ruin his off eye. It didn't seem to bother him much if he could figure out what was happening on his blind side. Everybody knew, if you came up on Hal from the blind side, you made a noise so he'd know you were there; it was polite, and helped keep from startling the old horse.

I went to the barn and got the bridle and saddle on Hal, which was not easy because I was still a kid and Hal was at least sixteen hands. I was proud that I was able to get that done without Daddy's help. It wasn't the first time because I'd been tacking up and riding Hal when I got home from school for some time by then. Getting up into the saddle also wasn't easy either. I knew from Daddy that the way to set your stirrup strap length was to measure it with your arm. The length of your arm is approximately the length needed for the stirrup strap. The problem is that kids and shorter people need shorter stirrup lengths which also results in the stirrup being further from the ground. That's not a major problem if someone else is around to give a boost or if there is a mounting stile around. I had neither and simply had to grab the saddle and clamber aboard as best I could.

I was excited the mission but a little nervous. It was my first time to ride out on the main road all by myself.

The ride itself was not eventful. There wasn't much traffic on the highway, and whenever a car came along, I'd guide Hal as far to the side as possible and stop as the car passed. It helped that Hal was blind on the right side so he didn't have any issue with the car startling him. The only odd thing was that one of the cars that came by was the county Sheriff, and he pulled to a stop.

Without getting out of his car, he asked, "What's your name boy?"

I was nervous as I answered, "Warren Gill."

"Is your horse for sale?"

"I don't know, sir. You'll have to check with Daddy."

"You Bill Gill's boy?"

"Yessir."

"What's your phone number."

"OL2431." I was proud for remembering as nervous as I was.

He wrote it down and drove off.

I took the horse on to Papa Gill's, cutting across Turkey Warren's place to avoid going through Petersburg. It was a short cut, and there was a gate near the creek that I knew about because sometimes, most likely on a Sunday afternoon, Peggy and I had used the gate when we'd go visit Turkey's older daughter, Margaret Ann, who was a little older than us, but was kin folks and always had some good stuff to play with.

When I got to Papa Gill's, he was watching for me and helped me take Hal's saddle and bridle off and told me to go in and visit with Mama Gill while he finished at the barn. Mama Gill always had candy around for a visiting grandchild, but it was usually something like jelly orange slices or powdered doughnuts, neither of which I cared for. She was so delighted to give them to me and was such a sweet, kind woman, that I always took whatever she gave and pretended like I loved it. Of course, she interpreted my words to mean that I wanted more, and she wouldn't hear of anything except that I have another piece. I think that day she wrapped some in foil to take home to share with Alan and Gloria.

It wasn't long before Papa Gill came along and told me he was ready to take me home. We got into his white Chrysler with the push-button gear shift and he drove me through Petersburg and out to our house. When we got there, we walked into the house to find Mama sitting next to the phone. She'd been crying, and appeared a little embarrassed about it, but was very glad to see us, especially me.

Papa Gill asked, "What's wrong, Carolyn?"

"I just hung up from talking to the Sheriff. He's gotta think I'm crazy."

"Why? What happened?" That was when I remembered the sheriff talking to me. I hadn't even told Papa Gill. I'd been enjoying the ride so much that talking to the sheriff hadn't seemed important.

Mama said, "When I answered the phone, all the poor man said was, 'This is the Sheriff. Was your boy, Warren, riding a horse out on the road on a big, sorrel horse?' "

Papa Gill couldn't help but grin as we both figured out what happened. When she got home and I wasn't there and Hal was gone, she would have thought I was out riding around the farm. She didn't know I was riding Hal to Papa Gill's, but she didn't need to. She instantly assumed the worse when the Sheriff asked his question.

Papa Gill asked, "Did you start crying into the phone?"

"No, I screamed! Then I cried. I couldn't help it.'

"Did he ever get to tell you why he called?

"Finally, when he got me calmed down. He wants to talk to you about buying Hal. He thinks Hal's a fine-looking horse."

"I guess I'd better call him. He leave a number?"

"Yessir." She gave it to him. As he took it, he said, "I'm sorry you had such a fright. I never even thought about the boy getting hurt."

By now she had added a little edge to her voice, "Mr. Warren, people drive so fast now that they've paved this road." She didn't openly criticize anything that Papa Gill, but she didn't have to. It just took those few tense words. Papa Gill could be harsh but he was he smart and could take a hint. I never got another call to ride Hal on the road to Papa Gill's again.

*****

Horses are the perfect animal. They take man's innate skill at dominating animals and give it high purpose. They magnify the results of man's labor, and often transcend the mundane to reach into the realm of beauty.

Power is the first gift of the horse. Immense musculature drives a marvelously engineered skeleton strapped together with steely ligaments and fed by marvelous lungs and a mighty heart. No other domesticated animal comes close to putting a package together like this, formed from birth for strength, speed and stamina.

Then there's the brain. Not big, not stupid, probably exactly right for fitting into a body designed for survival in a world where vast amounts of nutrient energy is bound within forages that can be had by brave browsers who can withstand the threat of predation with a combination of speed and strength.

Along comes man, comparatively lacking in strength, but with excess brains, and a partnership is formed. Man thinks - horses work. It is, and has been, an excellent relationship.

The first horses in the Cane Creek Valley were probably just horses, but they would have been the best that the settlers could afford, because carving a living out of the wilderness would be difficult at best, and practically impossible without good horses and mules.

The horses would be well cared for. One of the reasons for corn and oats was to feed the horses that helped make the crops. If the crops didn't make, the horses would weaken. A thin, weak mare won't raise as good a colt and may not breed.

A measure of a man is how he treats his horses. A man who lets his horses and mules go hungry is likely a no-count, or a drunk or both. A man who unjustly punishes his animals is likely to be mean in heart and will also beat his children and wife and is due a miserable end. Mules remember and often get even, as do children when they grow.

Winston Wiser was probably the best trainer of Tennessee Walking horses to ever live. He found and developed the great Midnight Sun, and when that horse's owner sent word that he no longer wanted Winston as trainer, he chased this great stallion out onto Highway 64 to fend for himself. Winston went on to beat Midnight Sun with Merry Go Boy in the 1945 Celebration. Those who witnessed this great show say it was one of the greatest contests ever. The crowd was divided. Papa Gill was for Go Boy. Daddy was for Sun. Fred Walker had been embarrassed in the Stallion preliminary and had been seen working Sun hard in the time between, but Winston Wiser and Go Boy took the night. The real story of these two stallions came when the power from Midnight Sun was genetically blended with the prancing showiness of Merry Go Boy to invent the modern Tennessee Walking Horse.

Winston Wiser was a great trainer with a poor reputation. There are many stories about his mistreatment of horses and people. Some of these stories have doubtlessly grown with the telling. It is tempting to include some here, but it might be better, and more entertaining, to encourage people to simply bring up the subject of Winston Wiser to any long-time horse show fan and spend the next few minutes hearing tales of the one of the inventors of the modern Walking Horse.

My favorite memories include the countless hours that I spent riding and showing Walking Horse ponies. I won some classes and lost more in small shows in places like Cornersville, Fayetteville, Pulaski and Lewisburg.

My first and possibly favorite, was a beautiful black gelding named Fury. He belonged to Deb Dodd, a farmer neighbor and friend of Papa Jepp's. That little pony had a nice little running walk and was fun to ride and show. I had to ride my bike about 3 miles to be able to work out with him, but that didn't seem like a big deal. Only problem was that I quickly outgrew him.

The next one I trained and showed was a larger, glass-eyed spotted pony that belonged to Joe Harris McAdams. Joe Harris was a large man with a big heart and a look that resulted in his nickname: Hog Head. His wife was a nurse, named Lola May, and she was a good match in both spirit and kindness. Two memories come to mind. First was my big win at the Fayetteville afternoon show. The pony had performed flawlessly, and I was so proud that I decided to show in the big show that night. I entered in the pony class, but soon realized I was out of my league. The 'ponies' in the night class were larger, professionally trained and just plain better. I came in last and will always remember that the winner that night was Pam Fitzgerald, who was not only an accomplished horseman but also beautiful, especially that Saturday night in Fayetteville.

The other memory was the dam of the pony I showed. She was thirty-six years old, and looked it, but she still won the costume class in the Petersburg Horse Show with Alan and Gloria driving behind her in a white buggy. Hog Head and Lola May were thrilled!

Mules

Before tractors, mules were the real workers. In some people's eyes, they still are. They keep easier than horses and give more work for their keep. They don't breed so there isn't any wasted time on foaling and nursing and all the foolishness that goes with it. A mule is made purely for work.

Mules are smarter than horses. A mule will be the first to figure the latch on the corn crib and will go in and eat what it needs and leave. The horse, who would've never figured the latch, will follow the mule in and eat till foundered.

Mules are too smart to work themselves to death. Anybody who has ever worked with mules has seen mules refuse to work if the job is too difficult or likely to cause the mule injury. The phrase 'stubborn as a mule' is likely rooted in this behavior, because a mule will get downright stubborn if it is too tired to carry on or is being asked to do something it doesn't want to do.

It is fascinating to work a mule in a place where it can demonstrate its ability to make decisions, like 'snaking' logs out of the woods. This is one job where mules are often at an advantage over tractors because mules fit between trees better, are more sure-footed, and have a brain.

It is obvious that pulling a large log through trees requires strength, but it also requires that a smart path be picked through the woods. A poor decision can result in anything from extra work to a log jammed between standing trees. The mule handler must do some of the thinking, but it helps if he can rely on the mule to make most of the choices. An experienced mule will invariably pick the best path, even if it isn't the most obvious.

I used Papa Jepp's old mule, Jack, many times for snaking logs out of the woods and found he almost always picked the path that was the safest, usually the shortest, and always the best overall route to the log pile. Jack understood better than I did how important it is to keep the log moving once it has started. Repeated starting and stopping wastes time and energy. It is far better to have a good path picked through the woods and get the log out as quickly and smoothly as possible.

The Colt Show

Horse trading was the engine of the development of the horse industry. Everybody needed horses and had some idea of the kind of horse they needed. Some people were skilled enough at understanding the relationship between utility and value that they could make a living at it. The same people, in today's world, are car or farm equipment dealers.

Early settlers had all kinds of horse, and traders brought more into the mix. Included were Morgans, Standardbreds, Saddlebreds, Thoroughbreds, Warm Bloods, ponies, trotters, pacers, mustangs, Belgians, Percherons and lots of just plain horses. Racing was evidently important in the early days, but so was farm work and pulling buggies.

The best horse for riding plantations wasn't necessarily the best for pulling stumps. Your friendly local horse trader could find what was best for your needs and he'd do if he could make a dollar or two.

The Civil War and Reconstruction turned the orderly world of Tennessee horses into a topsy-turvy mess. Instead of having specialized horses for specific needs, people had to try to find a horse that could do anything from plowing to working cattle. Mares had to work but also have fillies and colts and raise more horses and mules.

People struggled in those years, and their horses struggled with them, and there was a natural progression toward multi-use horses, but appreciation for beauty was never lost. If anything, it forced yet one more dimension onto the horse that was developing in the early twentieth century.

The competitive spirit, being as evident in rural folk as anyone else, has long found outlet in fairs and livestock shows. The people who 'invent' the classes of crops, vegetables, household goods, livestock, horses and mules base their groupings and guidelines on their knowledge of the realities of the agricultural world in their area. They offer competition for Burley Tobacco in those counties that grow Burley Tobacco and pick judges for the competition who are known to have the best understanding of the characteristics of that make an entry suitable or not. With tobacco, it may be numbers of higher quality leaves, length of leaves, color, and other factors that combine to give value to this plant.

Doubtlessly, the fair committees of the turn of the twentieth century struggled with how to construct classes for horses. Mules were probably easier because the value of mules is so causally linked to their utility. Horses have the utilitarian aspect, but this is usually eclipsed in competition by esthetics. The best mule is, and should be, that mule which is most likely to be able to go do the best day's work. The best horse is often that animal that looks the best as it stands or moves. Structural soundness, size, strength, (those things that support the ability to do work) are considered, but so are way of going (gait), beauty of the features (conformation) and ability to excite (showiness). The latter may be based on individual tastes, but, if enough people are drawn to a set of esthetic characteristics, standards are set and guidelines for judging can be developed. This may take some time, but, as standards become more uniform and horses are bred and trained to meet evolving guidelines, it becomes likely that classes and even entire shows can be built that promote competitive exhibition.

There can even be a living to be made. People pay money to watch exciting shows. People who work on the farm all week like to have a reason to go to town on Saturday night. Watching horses compete for prizes was evidently enough of a show to draw crowds to towns to pay enough money to fuel an industry.

The monetary 'fuel' for the horse show business is more than people paying to see horses, although that is part of it. The money also comes from people buying countless, delicious ham sandwiches and soft drinks and home-made pies at the shows. More money comes from people paying for horses and paying horse trainers to get the horses to move themselves in ways that excite crowds, catch the eye of the judges and win ribbons and prize money.

In the early part of the century, several shows became known for their ability to draw the best horses. The Wartrace Show was a great one. Belfast, Cornersville, Tullahoma, Eagleville and Pulaski also come to mind. Almost every small town had a horse show, but one of the best was in the Cane Creek Valley. It was called the 'Petersburg Colt Show and Community Fair' or simply the 'Colt Show.'

The Colt Show was more than a horse show. It was a gathering event. There were both horses and mules but there were also fruits and vegetables and cakes and pies and stalks of corn and tobacco. It was a community fair and a way for a rural people to gather in friendship, competition and fun.

The Colt Show was held in a creek bottom on the East side of Petersburg. This creek bottom was once surrounded on three sides by a large loop in the western tributary of the Cane Creek. Within that loop were the Colt Show buildings, an old house, a sawmill, and the Negro School.

The Chestnut Ridge Road (Highway 129) went through the Colt Show Grounds. This area was extremely flood prone, so a plan was developed and implemented by the W.P.A. to dig an alternate path for the creek, a straight run to allow quicker water movement during heavy rains. The result was better flood control for the bottom land with the added attraction of an elevated roadbed that allowed people to ride across the area even when the creek was up. Two dirt ramps allowed the crowds to drive down into the bottom land to attend the Colt Show.

The Colt Show structures were a wooden Grandstand connected by a rickety plank foot bridge to a barn-like exhibition building. The two-story exhibition building was filled during the Colt Show with all the wares that the local people could find, grow, cook, or make.

The stars of the show were the horses and mules. They were the best of the best. The mules, both red and black, were large and small and every size in between. They were sleek and powerful and drew crowds of admirers who appreciated the potential of these animals to pull almost any load worth pulling. Large heads and long ears may appear comical to the novice mule observer, but those who have spent time within the sphere of mule influence look beyond these surface oddities into the deep intelligence betrayed by large, brown eyes.

The mule handlers were as fascinating as the mules. It takes special people to devote their lives to animals that have such a unique combination of power and willfulness. These are hard people, or at least they work hard at making themselves appear as grizzled and tough as old wood. They put enough tobacco in their mouths to ensure its visibility, artfully cuss, and are given to wearing baggy overalls and big, well-worn hats. They also have skills with their animals that result in impressive displays of the ability of men to bend teams of animals to their will.

Mule men not only have animal skills; they also have plenty of people skills. If you ask a mule man why he likes mules, they'd each and every one have some batch of stories they could launch into about their beloved animals. I think the point is not how much they like their animals but how much they like to show their animals and talk about them to other folks. Sure, they like their mules, but they also know that a lot of people like mules and are attracted to folks with mules.

Think about it this way. Tractors essentially replaced mules in both my grandfathers' lifetimes. They hung around on a few farms for odd jobs like snaking logs and digging potatoes for several decades, but there are few jobs left for mules. Real, working farms do not use mules for any work at any time. The only exceptions I can think of are Mennonites and Amish who stay with old ways, including mules, for their own reasons. Yet mules haven't disappeared. In Tennessee, Walking Horse enthusiasts have a big horse show in Shelbyville, called the Celebration. Mule folks have a Celebration show at the same place only at a different time. They have Mule Day in Columbia, which seems to get bigger and better every year, at least if you believe press releases.

The mule show at the Tennessee State Fair, after some years of decline, is coming back. Hub Reese, one the most successful mule men in the United States, commented at a State Fair Planning meeting that mules and mule men were such a popular attraction, that the State Fair ought to pay those old boys extra for just being there. He may be right.

Back to the Colt Show. Some of the best Walking horses in the country showed at the Colt Show, as either weanlings, yearlings, or under saddle. A win at the Colt Show was as prestigious as any, until the Celebration, held in Shelbyville at about the same time, eclipsed the Petersburg event.

The Celebration started in 1939 and got bigger every year. The more the Celebration grew, the more the Colt Show waned. One year, in the early 50's, Winston Wiser got mad at the Celebration and decided to come to the Colt Show instead. That was the last time the Colt Show drew a decent crowd.

Uncle Allen

Uncle Allen Moore (we called him Uncle Allen, but he was really Daddy's first cousin) was always interested in both Walking Horses and mules. Uncle Allen was a smart man and a good ma, who earned his main living as a hardware salesman who traveled all over middle and west Tennessee. He always loved Walking Horses and never missed a chance to go to a horse show. He also loved children and always kept candy or gum in his pocket just in case he found a kid who he thought needed a sugar hit.

When we were little, he'd drop by at odd times and dramatically deny having any candy. The way he denied it, it was clear to us he had candy, but wanted us to wrestle him for it. In a wink, we were all over him, forcing him to the ground and searching his pockets until we found the hidden treat. In return, we'd gotten squeezed and tickled and had a glorious time.

As I got older, during the summer, he'd come by on a Friday or Saturday and take me to horse shows. He loved horse shows as much as any man I ever saw. He knew more people at the horse shows than anybody I ever knew. Once, at the Tullahoma Horse Show, he introduced me to Mr. French Brantley, the man who'd stood* Roan Allen and been one of the founders of the Breeders Association. He knew trainers and breeders and often had a horse or two in training.

*"stood" in this usage refers to keeping, or standing, a stallion for breeding purposes, usually for a stud fee.

Sometime around 1960, Uncle Allen bought an old dairy farm, just across the road from us. He'd kept a small herd of dairy cows, some beef cows, a few sheep and, by the time of his death, he had thirteen Walking horse mares and all of them were bred. The problem was that no one knew whether he'd chosen to breed them to a stallion or a jack (male breeding donkey). He usually made that decision based on how well walking horses were selling in relation to mules, but he had not left a record of what he'd bred to what.

Alan ran the estate sale and when it came time to sell the mares, all he could do was announce that he had some Walking Mares of excellent breeding, but some were bred to a stallion and some were bred to a jack. He'd be glad to tell the buyer the breeding when the mare foaled.

Shadow Lady and her band

(Note: This section was written several years ago by William Gill, IV)

When I was eight years old my grandfather had the smartest horse I've ever seen. Her name was Shadow Lady and she was the leader of Papa Gill's band of five horses. Shadow Lady was white, but she hadn't always been white. My father told me that she was dark colored, almost black, when she was born. She was a roan, though, which means she has some of white and black. She started life black with a little white and, by the time I knew her, she was almost solid white.

When I was born, she was about 15, so when I was eight, she was twenty-three. As old as she was, she got hurt constantly, but she could get out of anything. She was the leader of the horses on Papa's farm. There were five of them, counting Shadow Lady. There was Lucky Pride, Star Baby, Glowing Pride and Toby, the donkey.

Lucky Pride and Glowing Pride were Shadow Lady's get, or children. Lucky was a sorrel gelding, about five years old, and Glowing Pride was a bay, almost four. Another of Shadow Lady's offspring was a gray roan (almost black) gelding named Jim. When I was three or four and Jim was eight, Papa Gill sold him. Jim was something of a problem horse. He once bit Papa, so it was no surprise that he was sold.

Shadow Lady, Lucky Pride and Glowing Pride were all Tennessee Walking Horses, but Star Baby was not. She was half Morgan and half "who knows what." Star Baby was mine. She was given to me on Easter Day, 1990. Daddy bought her from Mr. Bill Hurst, who was Papa and Mama Gill's Methodist preacher.

Toby was a wild burro, or donkey, that Papa adopted from the Wild Horse and Donkey Ranch near Cross Plains. Toby was born in a desert in California and shipped to Cross Plains to find a new home. When Papa first brought him home, he wouldn't cross the creek, which left him alone whenever the other horses would wade the narrow crossing. It was almost a year before he learned the water wouldn't hurt him, and it took a drought to teach him. The drought caused the creek to dry up, and he started going across when it was dry. When the water came back, he just kept on crossing it.

Each horse in the small band had a different personality. Star Baby followed Shadow Lady the most faithfully, while Glowing Pride had her own mind. She still always followed Shadow Lady's lead, but she often had to do it differently. For example, if Shadow Lady decided that the band was going to the pond, she would lead the way there. Star Baby would be just behind her and Lucky would also closely follow. Glowing Pride would also go to the pond, but she would go around the barn to get there. Toby was attached to Glowing Pride, so he would probably also go around the barn.

Lucky was the smartest in one way; he could open gates better than any of the other horses. This was probably because he would do almost anything to get food. Papa had to put special latches on the gates to keep Lucky from opening them.

I usually only got to ride Star Baby on the weekends because of school. I was in the third grade. During the summer school break I got to ride more often. Sometimes Star Baby was hard to catch, maybe because Shadow Lady was so smart about avoiding the bridle. Shadow Lady was so old that she was hardly ever ridden, but she knew when she saw Papa or Daddy or me coming with a bridle that someone was going to have a good ride, and she started looking for a way to avoid it. The rest would follow her, so she would look for any escape that she could find. Sometimes it was a gate, sometimes just a far corner of the pasture. Fortunately for me, as smart a Shadow Lady was, Papa was smarter. We always caught them.

One way to catch them was with feed, usually corn. Papa often had corn in his barn that we would have to shuck and chop with a large knife. We couldn't fool Shadow Lady with feed, but Lucky was greedy about grain, and would almost always come to the feed bucket. After he started, Glowing Pride would usually come for her share and so would Star Baby. Usually, however, it was easier to just get in Papa's red truck and drive the horses to the barn.

After that we would catch whichever horses we wanted to ride. I always rode Star Baby. Papa or Daddy would ride either Glowing Pride or Lucky. Sometimes, if someone like J. Alan or Zack McDaniel was around, they would ride Shadow Lady. No one ever rode Toby because he wasn't broke to the saddle. Papa didn't get Toby for riding, but for keeping wild dogs and coyotes away from the cattle and for eating thistles. Maybe also for the fun of having him around. I doubt if he'd be very comfortable to ride, anyway.

Tacking Star Baby up could sometimes be a problem, mostly because she was too fat. Daddy or Papa would sometimes have to extend the girth on the saddle to get it to fit. For a long time, we had problems with her bridle, but Uncle Doyle (Meadows) gave me a bridle that his daughter, Mindy, was no longer using on her horses. This solved almost all our problems.

We often must work Papa's cows on horses. Lucky is a good cow horse. Daddy says he's got "pretty good cow sense for a Walking Horse." Glowing Pride and Star Baby are also good for getting the cows up. One time some of Daddy's heifers got out and we had to ride on three different farms to find them. We finally found them on Dan McGee's farm. The heifers didn't want to leave their new cow friends, but they had to go home. Daddy was on Lucky; Papa was riding Glowing Pride and I was on Star Baby. Those heifers didn't have a chance. We soon had them safely back on the farm.

When we rode, Papa's dog Lem followed close in front or behind us. Lem was a hunting dog and would hunt anything even though he was supposed to be a bird dog. One time we saw him come to point near a bush, and then a rabbit ran out the other side. Lem chased it, running faster than I ever saw before. Just as Lem was about to leap onto his prey, the rabbit turned and the puzzled dog gave up the chase.

When I rode Star Baby, it was usually a good sign if she went away from the barn in a cooperative mood. There is only one place that would bother her after we left the barn, which was going across the first ditch in Papa's largest pasture. Sometimes she didn't want to cross it, but she had to most of the time.

The only horse I've ever been thrown from was Star Baby. That was when there were a lot of horse flies (big, green flies that make horses nervous) around, and Star Baby was not in a good mood at all. She reared up and bucked when I was trying to make her leave the barn. She threw me off. My head hit one of Papa's gates, but I had my hard hat on that Papa and Mama Gill gave me, so I wasn't hurt. I wanted to get right back on, and Papa helped me do just that. I rode her for a while after that.

One other time, Papa and I rode across the creek onto the big bottom pasture. We rode for quite some time before Star Baby started "acting up." We couldn't figure out what was wrong. Finally, we discovered that her chin strap was broken. The chin strap is a small leather belt that goes under the horse's chin and secures the bit in place. After that we went back to the barn and fixed the bridle and finished our ride.

Sometimes we had to trim Shadow Lady's and Star Baby's hooves. They had both had a sickness in the past that had caused their hooves to grow incorrectly. The sickness was called "founder." There is no permanent cure for founder, but it's not too serious if the hooves are cared for regularly.

That's the story of Papa's Band of horses. It isn't the largest band and there aren't likely to be any world champions coming out of it any time soon. On the other hand, they are good horses with interesting personalities. They are fun to ride and almost as much fun to just be around.

Goats

The final farm enterprise we'll mention in this chapter is one of the most interesting of all...the goat. Curious, active explorer ruminants, they are smaller and a natural fit on the rocky slopes of the upland hills that border the Cane Creek Valley.

Cows graze but goats browse. They like nothing better than to stretch their necks, and even partially climb trees, to pick tender leaves with lips specially designed for selecting the best fodder available.

On the average, ten cows will give seven or eight calves a year worth about four or five hundred dollars apiece. In the same amount of time, on the same land, or even sorrier land, sixty does will give eighty kids worth some fifty or sixty dollars apiece. In some years, goats make more than cattle and, in some years, cattle make more than goats.

From a management standpoint, the main difference between goats and cattle is that goats are harder to fence in ... they climb over or through or even crawl under many of the same fences that will hold cattle quite adequately. This trait is a bother to farmers who like to keep their stock on their own property and a source of aggravation to the neighbors of farmers who have a lax attitude toward keeping goats fenced in.

The man who used to own the farm across from Daddy didn't worry much about where his goats were at a given moment. He knew they'd come home at night and he probably also knew that his goats probably would grow better if they got some of Daddy's clover into their bellies.

I thought having Mr. Welch's goats on Daddy's property was part of the natural order of things until one day, when Daddy saw Mr. Welch at Mr. Clifford Archer's Sinclair station in Petersburg. Daddy pulled the truck into the parking lot and stepped out without saying a word. Since Daddy was usually mild-mannered in his conversations with others, it was almost a shock as we realized that Daddy was upset with Mr. Welch over, of all things, the herd of goats that Daddy had just seen peacefully grazing on his bottom land.

If Alan and I were surprised, Mr. Welch was more than a little unnerved. Daddy asked him if he'd ever thought about fixing his fences and Mr. Welch started to stammer out some kind of reply when Daddy told him he didn't really care how he managed it, but the goats had better not be on his farm again. The tone in Daddy's voice led Mr. Welch to simply nod in acquiescent silence. That was all Daddy needed. He turned and left.

I wish I could report that we never saw the goats on our side of the road again. The most I can say is that we didn't see them as much for a few weeks. I suppose Mr. Welch penned them over on the east side of his farm for a while so they escaped more toward Mr. Gilbert's side or onto Uncle Allen's farm. Eventually, the goats started coming back to Daddy's farm as usual.

There's not a reason in the world why goats and cattle can't coexist on the same farm. Whenever universities have examined multi-species grazing, they have found increased efficiency. Adding one goat per cow to a farm probably doesn't have any effect at all on the number of cattle that can be carried. Cows tend to eat grass in the bottom lands while goats prefer to look for weed and tree leaves on the edges for forests and higher up the hillsides.

In other words, they eat different forages so the same farm can carry both without hurting stocking capacity.

Goats offer other advantages. They don't cost much, and they often have more than one kid at a time. They are small and easy to handle. They are a natural weed killer. They clean up weeds and brush without having to resort to chemicals. Maybe this is where the term 'brush goats' comes from.

Goat breeds are an interesting, if elusive, topic. By far the most common are the brush / bush goats. It bothers some people to refer to goats by this nebulous terminology, so they point out that at least part of these animal's ancestry is derived from the so-called Spanish Goats that filtered into Tennessee from Texas and Arkansas, etc., and has been fused over the decades with assorted milk breeds, mostly Nubians.

In the late 1800's, the Stiff-legged goats were introduced into Marshall County. This breed's falling-over trait makes them interesting, but they also have distinctive features, like a compact face and a stocky, meaty frame. Many of the goat herds in the Cane Creek area had at least some Nervous Goats.

Goats are a good size for a barbeque. A fifty or sixty pound kid can be a great main dish at a family fourth of July Celebration, and Papa Gill liked to do just that. Don't try it with an old billy, though (a billy is a male goat). One time, somebody gave Uncle Allen an old billy, and he dressed it on the farm and cut it up and brought it home for Aunt Mary George to cook. She noticed it had a hint of that distinctive billy goat odor and suggested it may not be worth cooking. Uncle Allen assured her that it was a young billy so the off odor would go away if she cooked it long enough. She wasn't convinced and asked, "Are you sure?"

"Of course, I'm sure. This is exactly the way Uncle Warren likes 'em. Aunt Cooper just boils 'em til the meat falls off the bone. Put a little sauce on it and it'll be delicious." After he convinced her to try it, he left for the farm.

When he got back several hours later, Uncle Allen was met at the back door by an angry wife. She'd boiled that billy goat for hours and the smell had only gotten stronger. The whole house had become steamed with the unique and pervasive odor of male goat. The longer she cooked it the stronger the smell. The stronger the smell, the madder Aunt George became. She hated to give up, but the smell wasn't the only problem....the meat showed absolutely no sign of falling off the bone. Quite the opposite. It looked more like it was being transformed into some form of gray rubbery tissue that was more tightly bound to the bone than ever.

When Uncle Allen walked in, he smelled the problem only a split second before her words slammed him like a brick. Her orders were, "Allen Moore, take that damned goat and the pot it's in and bury every bit of it. Bury it deep! I don't ever want to see or smell any of part of it ever. Do you understand? And don't ever even mention goat in this house again!"

Hard Times

Time heals. It is always easier to remember the good times, the happy moments, the lazy Sundays. It is not always fun to remember how difficult things could be during hard times in the Cane Creek valley.

The Great Depression was hard for the farm economy, but the farmers around the Cane, with their cows and hogs and corn and tobacco could always find a few dollars, and they all had big gardens and hunting to supplement the table. They'd read about farmers being driven from their land and knew that, as bad as the Depression could be, it was worse in other places.

A few times, great clouds of fine dust blew over the Cane Creek valley. It was from the Oklahoma dust bowl. The dust blotted out the sun and forced people to flee into their homes, but it even followed them inside. Efforts were made to seal windows and doors, but the dust still got in. The dust was a problem, but there had to be a certain element of relief by comparing the discomfort of a dust cloud with the total misery of the people who were living in the places from which the cloud originated. Oklahoma is a long way from the Cane Creek valley.

The Great Depression

I remember Papa Gill (Senior) reminiscing about the Great Depression and World War II. He was running a hardware store in those days and carried many a debt for farmers who had a hard time making it through. He was also deeply involved in the Production Credit Association, a cooperatively owned farm credit organization. We have a tape of him that was professionally developed by the Production Credit Association during the war. The idea was that Papa Gill, playing a hard-working but struggling farmer, was writing a letter to his son (Daddy) about how farmers were supporting the farm effort, overcoming the Depression, and helping win the war at the same time. The end of the tape was a segment in which he gave a speech encouraging other farmers to borrow wisely and produce more for the war effort. It was a good strategy that helped a lot of farmers through difficult times.

Papa Gill's role in the Depression was probably more public than Papa Jepp's, but you wouldn't believe it by the stories. The one that I remember best wasn't so much a story as a lesson in economy. In fact, after I'd gotten it a few times, I became a little afraid of the 'Great Depression' lesson, in which I was to learn to appreciate the value of whatever object was under his scrutiny now.

It was still Spring but getting toward Summer. The days were getting longer and warmer but were not yet uncomfortable. Mama had just told me that Papa Jepp was coming over to take us fishing.

When he arrived I was already outside and excited about the fishing trip. He was walking under the Maple tree in our back yard where Mama usually kept lawn chairs for visiting with company. As he sat down, he looked to the ground and saw a penny which he scooped up with a smile, as if he'd found a treasure. My response was both smart alecky and dumb because I said, "I knew that penny was there."

He looked at me in disbelief that I would make such an admission, but I think I also saw he same look he got just before he snatched a sucker. He asked, "You knew it and didn't pick it up?"

I was in trouble. I tried to minimize my error, saying, "I thought I'd get it when I needed it."

My comment didn't help. It just added an extra element about 'putting money in the bank' to the 'Great Depression' speech. This one took a good twenty minutes and covered plenty of facts about how important pennies were in staving off starvation when jobs were scarce and farmers were struggling. I learned that tobacco was as low as three pennies a pound and that children would scramble in the dirt to get hold of a penny, and I was just walking by it like it didn't amount to anything.

He was still going strong when Mama came by and heard what he was lecturing me about. I was pleasantly surprised when I saw her roll her eyes (behind his back) as a signal that she understood my predicament, but I didn't dare smile. She first tried to head him off with a defensive remark about how the Depression was over, but that only made him think of how we could go back into one if people forgot the value of money and the hard work it took to make it. Then Mama changed her tack, reminding him that he'd planned to go fishing. Fishing was the one thing that could always be counted on to divert Papa Jepp's attention.

The Big Barn Burns

One year, I was around six, Alan was four, and Gloria was still in diapers, was the most difficult. We had one big barn, it was old and very large. I don't know if I truly remember it or not, but I sort of think I do.

I remember the night it burned. There was a lot of activity, with trucks riding by late at night. I'd gone to bed, but the sounds of people shouting woke, me and I went out to see what was going on. The barn was about three quarters of a mile from the house. When I went out the back door, I expected the usual darkness, but the sky was full of a fiery, smoky glow. It was in the direction of the big barn. I watched it for a while, unsure of what was going on, but I knew it wasn't normal. Mama was scurrying as much as the rest until she saw me. She swept me up and took me back into the house and told me to stay in bed.

The barn burning destroyed all of Daddy's tobacco and most of his hay. He didn't have much milking equipment, but what he had was gone. The tobacco was to have been most of his income for the year and the hay was supposed to have kept his milk cows going for the winter.

The farm was far from paid for, so Daddy had to make payments. He also had to find the money to keep food on the table and to pay the doctor when kids got sick. There were light bills and phone bills and plenty of other expenses. In other words, he had to find money to keep us going. He found a job as a welder of concrete truck equipment in Shelbyville for a year or so.

It was still tough. I remember Papa Gill coming by one afternoon and talking to Mama and Daddy on the back porch. I have no idea what the conversation was about, but Mama was upset. I was playing out back, close enough to hear Mama's distress. Then she opened the screen door and called me over. She held my shoulder and pointed at my jeans as she told Papa Gill, "Look how I've had to patch his britches. We don't have money for clothes!" I'll always remember the sad look on Papa Gill's face.

We never missed a meal. We had old cars and trucks and made farm equipment last. Daddy sold a working Model A truck for twenty-five dollars. He sold all his goats, about a hundred, for fifty dollars. Not fifty dollars each. One hundred goats for fifty dollars! Fifty cents apiece! I'm sure it was tough for Mama and Daddy, but we were always warm enough in the winter, Mama's cooking was always delicious and filling, we always got to church, and were almost always laughing and joking.

When the Soviet Union fired Sputnik, things started getting better. I was young, but distinctly remember all the scary press coverage. The Weekly Reader was saturated with reports about going to space. President Eisenhower was determined that the United States wasn't going to lose out in the race for space. That triggered massive hiring at the Redstone Arsenal in Huntsville, which is only about forty-five miles away.

Aunt Mary Neil got a job down there and soon started dating an Army Engineer named Bob Lindstrom. He was from Chicago, which made him a Yankee and a little suspect, but everyone liked him. Before we knew it, she'd married this intelligent and good man and was living in Huntsville. Uncle Bob had left the Army to take a management position with the newly formed National Aeronautics and Space Administration – NASA.

We heard that they were paying an unbelievable five dollars an hour at the Redstone Arsenal, and Uncle Bob was important enough that he was able to help Daddy get a job. Then, with a little help from Daddy, Uncle Edward applied and got on.

Things even got more secure when President Kennedy called for an all-out national effort to put a man on the moon within ten years. I don't know if the race to the moon was as important to others as it was to us, but it was life-changing for the world around Huntsville, Alabama. It pumped money into the region and brought people from around the world to tiny Huntsville. Huntsville turned from a sleepy little cotton-economy town into a booming city.

Uncle Edward soon moved his family to Huntsville and, with Aunt Mary Neil already living there, Mama started wondering if we should move there. I was petrified. I couldn't imagine living anywhere but on the farm, in the big house by the Little Cane Creek.

More than once we went to Huntsville and spent a Sunday afternoon riding around and looking at houses. Mama knew I hated looking at houses, so she'd often drip us off at Aunt Mary Neil's who would take us swimming in a nearby community pool. Gloria and Alan loved the pool, but I didn't. (More on this in the Cousins section.

I could understand why Mama might want to live in Huntsville, but my impression was that Daddy didn't like the idea any more than I did. All I could hope for was that Daddy was playing a stalling game until Mama decided we could find a way to stay where we belonged. That's how it worked out, whether Daddy had anything to say about it or not. Maybe it had to do with Mama getting a job as a teacher.

Daddy spent the next twenty years working at the Redstone Arsenal during the week and running a four-hundred-acre farm in the evenings and on weekends. The drive was forty-five miles, one way, and he car-pooled with three others to save a little money and make it more bearable. I think he may have napped when he wasn't driving, because he'd come home and be ready to work for a couple of hours if there was enough light. Even if there wasn't enough light, he'd find something he could do.

It was amazing to see how much work Daddy could do on a Saturday. He'd also get as much work out of Alan and me as he could, but it was nothing compared to what he accomplished. He raised crops of tobacco with his Saturday / after-work labor system while others struggled to raise smaller crops by working at it all week. He kept the beef herd going, and fences mended. He built a new barn and, a few years later, another barn.

Daddy cleared more land and kept the land clear by bush-hogging every cleared acre every year. He even bought ninety more acres that joined us on the south.

Bush-hogging

Bush-hogging is an interesting part of the art of farming - I say art, because I rarely hear it included in the scientific approaches to farm management, but most farmers know that farming without a bush-hog would be a lot more difficult. A bush-hog is a heavily built rotary mower that can mow a wide range of plant life, from pasture clipping to whacking down small trees. Out west, they call bush-hogs 'shredders,' which is a more accurate name. Bush Hog descriptive of what they do. The original bush-hog, built by the Bush Hog company, advertised that it could cut down anything that the tractor could drive over. As far as I know, that was true. I've seen Daddy use a bush-hog to cut trees that were big enough that the front of the tractor was raised from the ground as he drove them down. When the trees got under the mower, there's was an awful noise, but the young trees would disappear into a hail of wood chips and only the ragged stump remained.

Bush-hogging was easy enough on the level to rolling bottomland but has always been a challenge on our steeply sloping hills and over rocky terrain. Daddy has been through several bush-hogs over the years as he tried to win the battle against the creeping forest that is always threatening to take over. Nature seems to resent the fact that farmers want to keep a little open space between the forest and the Little Cane Creek. If job and family put demands on time, or the farmer gets sick or injured for even few months, the forest starts attempting to take back the land.

If the land isn't regularly mowed or sprayed, there many different types of weeds, bushes and brambles that grow, first along the edges, but eventually everywhere that pasture has been established with such hard work and expense. The first step in the takeover is often a weakening of the Fescue stand, maybe due to drought or simply failing to keep the land limed and fertilized. Fescue competes very well, if it has half a chance, but if it gets stressed out, you start seeing more weeds like iron weed, pigweed, cockleburs, and horse nettle; then more of sorry grasses like greasy-grass and broom sedge (commonly called sage grass). Bermudagrass and Crabgrass also put out their complex root systems and take over whole sections of pasture. Then come the Blackberries and Multi-Flora rose which can take over a field in only a couple of years of inattention. Buck Bush (AKA Buck Brush) and thistles also do their part, with a little help from trumpet vine, burdock, and a few dozen others.

If you let the field go longer without attention, Cedar seedlings start growing into trees. Sumac and black locust sprouts get big fast. Honey locust and Bodock also grow fast and have nasty stickers that makes cutting them a painful ordeal.

Fescue Hay

One of the yearly challenges was to get enough hay put up to keep the herd going all winter. The first challenge in getting a good hay crop is to have a good forage base. Kentucky 31 Tall Fescue has been a godsend in this regard. Fescue is a common grass around the world, but the Kentucky 31 variety was discovered by Professor E. N. Fergus on Mr. W. M. Suiter's farm in Menifee County, Kentucky, in 1931. This hardy grass which grew so well in cold weather and on all kinds of land, was studied by the University of Kentucky for years before it was released.

One reason Kentucky 31 Fescue was studied for so long was that it was a 'mixed bag' forage. It was a weak seedling that was sometimes puny at the beginning of its life, but after it took hold it was one of the most tenacious forage plants that had ever been seen in the mid-south. Animals didn't like it as much as Orchardgrass or Bluegrass, but Orchardgrass didn't persist in pastures or hayfields and had to be replanted every three to five years and Bluegrass is finicky about where it will thrive and doesn't grow enough quantity to make good hay.

Another problem with Fescue is that cattle are not as productive on it, especially after it matures. In the forties and fifties, when the seed was being sowed across Kentucky, Tennessee, and other states, farmers took to it because, after it caught it stayed, even on steep hills. Calves grazing pure Fescue didn't grow as well, and cows didn't slick off (lose hair in the Spring) and get fat enough to breed back. In the late seventies, ag scientists realized that a fungus that lived inside the Fescue was putting out alkaloids that were toxic enough to hurt production. The fungus grows up inside the plant as the plant grows. The fungus can't be seen without a microscope. In the Winter, the fungus hides in the dormant roots, and in the Spring, the plant grows so fast that the fungus can't keep up. As Summer approaches and the plant matures, the fungus spreads within the plant. By the time the Fescue plant is in its reproductive phase, the fungus has crept up the stem and has thoroughly saturated the plant, especially the seeds. When farmers sow Kentucky 31 Fescue, they are sowing both Fescue and the fungus.

Somebody thought to go back the original farm in Menifee County, Kentucky, and found the fungus was in the original field and were passed from the seeds harvested there until it covered practically every field in the Cane Creek valley and millions of other acres in Tennessee and the mid-south region.

Fescue was probably one of the main reasons that Daddy could keep his beef operation going. He sowed it on most of the pasture and hay fields. He found he could get practically all his hay up during a few weeks of hard work in May and early June. Fescue harvested at this time was still tender enough to make nutritious hay and yielded enough to fill the barns. In the fifties and sixties, we didn't know the fungus was inside the plant, so nobody knew we were harvesting the hay when the fungus was least likely to have progressed up the stem and into the leaves and seeds of the plant. Without knowing the fungus was there, we nevertheless were learning to manage around it.

Farmers have known for centuries that legumes like clover and lespedeza are good forages for livestock. It didn't take farmers long to discover, with the help of university research and Extension agents, that adding Ladino Clover, Red Clover and Kobe Lespedeza to fescue pastures made the forage more nutritious. Now, we know that clover ameliorates the damage to performance that the fungus causes. Yet again we were managing around the fungus without knowing it.

After the fungus was discovered, scientists quickly discovered ways to get rid of it. Trouble was, when they got rid of the fungus, the plant wasn't as strong. Obviously, the fungus needs the plant to survive, and now we know that the plant also needs the fungus. It is classic symbiosis. It may be a more complicated relationship than simple symbiosis. The plant needs the fungus to thrive, and cattle need the plant because no other plant 'fits' the kind of land and part-time management that Daddy has done since the fifties. The land that supports the cattle needs a persistent grass with a complex root mat to prevent erosion. Fungus-infested fescue did the job.

Besides giving abundant hay, Fescue gave another gift that helped Daddy thrive in the Beef business - Fall growth. No other grass grows in the Fall like Fescue. When the weather starts getting cooler in the Fall, the plant senses the coming of winter and puts its energy into getting ready for the freeze. Its deep roots pull nutrition from the soil which lets leaves put their photosynthesis factories to work to make sugar. The sugar-fed leaves grow slick and green and are full of sugar and protein. In the spring, a large portion of energy goes toward reproduction, but in the fall the plant throws itself into its cold preservation strategy. There is no seed production in the fall.

Fescue is so good at building nutrition reserves that it builds at least ten-fold more than it needs. The plant scientists refer to the ability of plants to take up and store extra nutrients as luxury consumption. This extra boost, if managed, can keep a herd of cows going all Fall and well into the winter.

The discoveries that Daddy was making in growing his beef herd were being made on dozens of farms in the Cane Creek Valley and on thousands of farms in Tennessee and thousands more in Kentucky and thousands and thousands more in Alabama, Mississippi, Arkansas, Georgia, the Carolinas, Virginia, and Missouri. The cow-calf business thrived on Papa's farm and it thrived in the region.

One of the cows in our current herd.

It is a cascade of symbiosis. The plant needs the fungus. The cattle and the land need the plant. The farmer needs the cattle to make farming pay and a stable land base for long-term investment. The cattle industry needs farmers like Daddy who know how to turn land and grass into good beef calves and are willing to sell them cheap enough that the industry can take them and profitably turn them into a basic and integral part of feeding our country and the world.

As neat as this story sounds, it should be remembered that the existence of the beef herd is evidence of disappointment, struggle, and hard times. As tough as it was, the burning of the big barn likely only hastened the changes that were inevitably coming. In the sixties and seventies, countless small farm operators were forced into taking jobs in town. Milk barns were turned into farm storage or simply allowed to fall into decay. Hog farms disappeared, at first only a few, but as the twentieth century ended, it became difficult to find a hog farm. The sheep industry, once thriving, quietly faded into a remnant, nursed along because it offers one major benefit: sheep are a great youth project for 4-H and FFA Clubs.

Economists can explain the changes that occurred in the make-up of agriculture in the Cane Creek Valley and throughout the region. These lost commodities represent shattered dreams. People bought farms thinking they could work hard and live a good life. Instead, many of them worked hard and lost everything. Others lost their dream but managed to save the farms by converting to beef cow-calf production and getting a job in town.

'Beef cattle cannot pay for land.' With a few notable exceptions, this statement has been true for many years and remains true today. When you see a beef farm, you are seeing a farm that was paid for by something other than the cattle. It may have been paid for by other commodities when the other commodities were more viable. The farm has more likely been paid for by off-farm income.

Daddy's changes may have started with the burning barn but didn't end there. After the big barn burned, and Daddy got a job, he decided to start a beef herd by turning an Angus bull with the dairy cows. These dairy cows were mostly grade Jerseys with a couple of milking Shorthorns. The Jersey breed was great for small milk herds. They don't give as milk as much as Holsteins, but their milk is richer because it higher in butterfat. Butterfat is important because it is necessary for butter and cheese so milk that is higher in butterfat is worth more. Jerseys are also smaller than other breeds, so they are easier to handle and they eat less hay and supplement.

Little cows that give rich milk on less feed are great for a small dairy herd. It is questionable that Jerseys would be the ideal basis for a beef herd, but they were what Daddy had, and all that he could probably afford at the time. It turns out they did all right. By breeding them to the best bulls he could find (and afford), he got black calves out of cows that gave a lot of milk to make the calves grow fast. These calves may not have been ideal beef calves, but they were worth enough to the growing beef industry to pump income into the recovering farm operation. This allowed buying better bulls and upgrading the herd.

Daddy not only sold dozens of good calves every year. We also got to eat these half Angus half Jersey animals. Daddy was doing well enough in his work that we were able to start buying things like other people had, like a TV and a better car and a freezer. To fill the freezer, we needed beef and Daddy had a farm full of beef. Daddy was too smart to put any of his best calves in the freezer - they were worth too many dollars which were still rare and precious. Instead, we usually got to eat an animal that was on the lower end of the quality range, or a problem for some other reason. We had plenty of meat, but the steaks we got were often mis-steaks!!

One time, a heifer got out and was hit by a car. It screwed up the car and killed the heifer, but Daddy cut her throat and we took her to Pud Moore's in Fayetteville. Pud turned that poor beast into hamburger and a few steaks and roasts. This was pure grass-fed beef - not exactly melt-in-your-mouth delicious, but Mama figured ways to cook it and make it edible. It was pretty good most of the time.

We got our best freezer beef when Daddy picked a one or two cull-quality young heifers that were fed corn and supplement for a few weeks. University of Tennessee research has shown two relevant facts. One, taste panel research has shown that Jerseys yield the tastiest beef of all the breeds tested. I think the half Angus, half Jersey calves we ate in those years were genetically favored to give us good meat, and it was good.

The other fact revealed by UT research is that calves need to be fed at least 50 days for the meat to be rid of the less delectable blends of fatty acids and other chemicals that come from grass. In other words, the corn-fed flavor comes from at least a couple of months of being fed corn. Daddy's best beef came after he started feeding them for longer periods.

Not all the original herd was Jersey. There were also a few Milking Shorthorns. These were bigger cows and more naturally adapted to beef production. The Shorthorn breed had separated into a beef breed and a milking breed many years back, but the milk breed hadn't been widely successful. Neither had the beef breed, but it has continued through the years to offer enough advantages that it has maintained viability. It got a boost in recent years when the association allowed beef Shorthorn breeders to use Milking Shorthorn genetics in their breeding. This allowed them to increase both the frame size and the milking potential of their cows. It has seemed to help, but so has clandestine infusion of other breeds that have improved the meatiness of the breed without changing the unique range of white-red-roan color pattern.

One of the original Milking Shorthorn cows was the largest cow in the milk herd. She was also, by far, the most irascible. Daddy called this kind of cow "high-headed" (crazy, nervous, hard to work with) because when they saw a person coming their heads went high, their eyes opened wide and they started looking for a way to cause trouble.

That ole Shorthorn was always the hardest to milk, the most reluctant to go in the barn and the most likely to kick. She would have never lasted in the milk herd, but she was in luck. She was young enough and beefy enough that Daddy kept her as a beef cow.

She quickly adapted to the role, becoming the most contrary beef cow in the herd. When all the other cows came to the barn, she stayed outside until there was no sign of a human around. Then she'd go in and chase the other cows away from her eating place. When it came time to work the cows, she was always difficult to trap. She'd never come to a call, like the others, but had to be chased into the pen. When she got chased, she got excited and tried to break out by crashing through the pen fence. She was big enough that she could and did succeed, which forced Daddy to build stronger fences.

When we got her in the chute, she'd resist letting her head be caught, and when we finally caught her, she'd twist and bawl like we were torturing her. This got all the other cows nervous and made the whole bunch harder to work.

She was smart. She'd watch us from the moment we appeared and knew when we were getting ready to do anything that might affect her day. If it was time to wean calves, for example, she'd see us coming to get the herd up to figure out what we were about. If she could, she'd instantly take her calf and head for the most remote part of the farm.

So why did Daddy keep her? The answer was simple. She had the best calf in the herd year after year. It was always the thickest, meatiest and fastest growing. If it was bull calf, we'd 'steer' it and it would sell for more than any other calf. It was always questionable whether the quality of the calf was worth the hassle of keeping her, but she somehow managed to stay around.

Keeping high-headed cows is a common phenomenon. Practically every beef herd has one or two cows that are problems. Usually, they are intelligent. They are the ones that are most likely to find a weak place in the fence and break into the wheat pasture before they are supposed to. They understand that when a stranger is around, it could mean there is something going on that could affect them, so they look for ways to escape.

The smart ones are often the boss cows. Like Daddy's Milking Shorthorn, they use their intelligence and strength to make certain they get the best location at the hay rack, the best pasture and the coolest place under the shade tree. Maybe being boss makes them naturally a little mean.

Most cattle folks suspect that this bossy / mean trait is genetic. Daddy saved the heifer calves out of the old Milking Shorthorn cow because they were the best-looking heifers. They invariably became like their mother, bossy, mean and intelligent and they gave the best calves in the herd. This held true for at least four generations.

Behavior is either inborn or learned. If it is inborn, there is a gene, or some combination of genes, that makes cows intelligent / mean and it is linked with beef quality. If it is learned, then that old Milking Shorthorn did a great job in teaching lessons on becoming a contrarian cow.

Bangs

Twenty years of breeding to good black bulls turned Daddy's small milking herd into a productive commercial cow-calf herd. He'd built a new barn for hay storage and had put in a decent head gate and chute for working the cattle. His pastures were in decent shape and he'd begun figuring out how to manage the fescue to the best advantage. Things were going well, so fate decreed that we were due another set-back. It was time for Bangs to strike

Bangs it the common name for Brucellosis. Dr. Bangs identified the bacteria that causes it in 1897 so his name got tangled with this serious disease. It is Undulant Fever in humans. It had been around forever, always a sneaky, pesky disease that robbed profits on cattle by causing abortions. The fact that it was linked to a human disease was worrisome, but pasteurization had made it less dangerous.

Nevertheless, Bangs was bad enough that when a vaccine was developed, it occurred to some veterinarians at the USDA that it might be possible to rid the country of this disease. The problem is that the vaccine itself wasn't a hundred percent effective. Since the vaccine was only partially effective, getting rid of the disease would take a gigantic coordinated program that included vaccinating, testing and culling infected animals. In many herds with serious levels of infection, the entire herd would have to be eradicated. Money was allocated to pay farmers for culling cows. They got the market value of the culled cow plus a hundred dollars or so -- an amount that would not really satisfy the farmer or make him want to sell, but enough to make it less painful.

Herds were selected for testing when infected animals were discovered, often when cull cows were sold at the local stockyard. When a herd had animals that tested positive, indicating the presence of Bangs, all neighboring farms' cows were tested. That's what happened with Daddy. A neighbor's cows tested positive, so they had to test his herd. Guess what? Some of Daddy's tested positive. He culled the positive-testing cows and hoped it was over.

It wasn't over. The government testing man, Richard, was a big bear of a guy who was as friendly and nice as you will ever meet, and the last person in the world who wanted to bring bad news. But bring it he did. When they tested again, they found two more positives. That meant those two had to go and the whole herd had to be tested again.

Every time the cows got tested, they became warier and harder to work.

During the next testing, the government man and Daddy discussed options. Richard told him that his cows might continue to have positives every time they got tested. It could be due to the presence of a low level of the disease, or simply because they'd had a mild form some time in their lives and were still testing positive. Sometimes cows tested positive because they'd been vaccinated. The vaccine created 'false positives,' but the cows tested positive nonetheless, and would have to be culled. The only way to be truly rid of the problem was to 'Bang out,' which meant sell the herd for slaughter. Richard didn't encourage any option, but you could tell he knew it would be smart to take the support money. He said the decision was Daddy's.

It is never easy to send productive animals down the road. It is much harder to simply flush a herd that has taken years to build from a bunch of low grade Jerseys. Sometimes you've got to cut your losses. Daddy made the hard decision. It took several trucks to haul away the herd.

Loading them out was physically difficult. The old loading chute wasn't up for loading that many cows at once, and it had to be repaired several times during the process. The cows sensed that something awful was happening, and they made it even more difficult by resisting at every step of the way. Maybe that was good. By the time the last cow was loaded and gone, everyone was thinking they weren't such good cows after all. The cows sold for a good price, and Daddy got an extra hundred dollars a head from the government. Shortly after they sold, the cattle market crashed. There are always ups and downs in the cattle market, but very few crashes. This was a real crash: a forty-cent drop in price per pound. It happened quickly and made everyone in the cattle business worry about their future.

Daddy took the money he made on the cows and bought a load of steers at the bottom of the market. He was probably as nervous about investing in cattle as anyone, but he had grass and a barn full of hay he'd planned to feed to the cows. He also had money he needed to put back in cattle or pay a lot of extra taxes, so he used all that money to buy steers.

This idea of buying steers for growing out has a name, although it's a little clunky. It is called backgrounding and is simply the growing of weaned calves to heavier weights. Usually, this 'hardens' the calves and gives them a 'background' that makes them more valuable to the western feed yards. In some parts of the country they call this practice 'stockering,' but I don't see that this word is much less clunky than backgrounding. Maybe someone should make up a special word for it.

Daddy wanted to background steers because they were not likely carriers of Bangs and he needed to leave the farm free of cows for a year or two to allow the threat of brucellosis to pass.

The crash of the fall of '78 allowed Daddy top buy the steers for a low price. The depression in the cattle market only lasted a few months before the prices started strengthening. By the time he sold the steer the next summer, prices were up. He made more money on that group of cattle than on any others of his life. Sometimes hard times turn into good times.

He backgrounded another set of steers and did all right on them, and then he started buying cows. This was the start of a new cow herd which, through the years, has been turned into the herd we have today.

More good news: the USDA effort to rid the state of Brucellosis worked. Tennessee is now officially Bangs-Free. The entire country is practically free of this insidious disease.

Trees

Trees line the banks of the Cane Creek, giving shade that cools the water and providing permanence to the homes of the creatures of the creek bank. Creek willows and sycamores love to have their roots bathed in water and seek to live their lives right at the juncture of water and earth. This is a good deal for the creek bank because the growing roots stabilize the soil and firm the path of the stream.

Trees along the Little Cane.

Box Elders and Elms thrive in the riparian zone, as do Walnuts and the ever-present Hackberries. Privet and multiflora rose fill in where low growers need to be. I suppose the Cane Creek once had a lot a cane, but there's not so much now and there hasn't been for a while.

The combination of trees and water create an environment that draws animals. The trees are always alive with squirrels and birds.

Cows and horses spend their resting hours in favorite spots along the creek bank. If hogs can get to the creek, they'll root and dig until everything is a mess, so most farmers keep the hogs closer to the barn.

Further from the creek, as the soil becomes shallower but better drained, there appear to be more of the Maples and Oaks. White Oaks, Black Oaks and Red Oaks all proliferate. Most of the Maples are Sugar Maples, and there aren't many. Jack Daniel's uses a lot of Maple for making charcoal to filter their whiskey, and they've paid good money for a long time for good Maple logs, so a lot of the Maple trees have disappeared into Lynchburg from the counties around lower middle Tennessee.

Maple tree near our home.

Walnuts

Black Walnuts also do very well in the forests along the Cane. Walnuts are esteemed for the beauty of their wood, but the nuts have a great flavor and some value. When the nuts start falling in late September and October, money is there for the taking. We'd gather five-gallon buckets and tow (burlap) sacks and head for our favorite Walnut trees. The best trees were the ones that gave the huge nuts. Some trees, probably those grown on the poorest soil, tended to have little nuts with thin, dry skins. It took forever to fill a sack with these little fellers.

The big nuts were a joy to pick up. First, we'd try to get as many of the nuts off the tree as possible. If we could, we'd climb up and try to shake them off. Sometimes, we'd use a long pole to knock them loose or throw sticks up amongst 'em. Alan was, by far, the best at knocking walnuts off trees by throwing something at 'em. It was usually a large stick which could whack into a high branch and loosen a dozen or more fat walnuts at a time. Alan probably liked the throwing part more than the picking up, which worked out fine because the rest of us could do the picking up.

One time, Alan and I decided we could pick up Walnuts fastest by parking the old black Ford Pickup close to the tree and throwing the walnuts in the bed. We still used buckets part of the time, but we avoided the use of burlap bags. We filled the bed of the truck about two thirds full and decided to cross some rocky places above the pond to get to better walnuts. The truck was so full that it was riding down on the back springs. I'd only been driving for a short while and Alan didn't have his license yet. As we were trying to thread through the rough ground, we must have miscalculated and hit a sharp rock. Boom!

One of the back tires blew out. What a mess. As I remember it, we tried to jack up the truck for a couple of hours and failed and finally shoveled a bunch of the walnuts out and still couldn't get it raised. As usual, we had to wait for Daddy to come up with a solution, which he did. He also helped us bag out the walnuts and pretty much saved the day.

The years we became the most serious about picking up walnuts was when Braden and Annie Lee Wilson were living in the farmhouse closest to us, to the east. I mentioned before that they moved there when they sold their grocery store in Lewisburg and decided to retire to a farm that was for sale on the Little Cane Creek. Annie Lee's mother also lived with them. We called her Big Mama because she was very heavy and a delightfully happy, wise, and friendly lady. Their daughter, Bradeen, visited often enough that we also counted her as part of the family. Not long after they moved to the farm, Bradeen adopted a daughter, Barbara, who was a year ahead of me in school but who had such energy and was so delighted to do things with us, like picking up walnuts and riding horses, that she became our best friend in those years.

Mr. Wilson wasn't much for picking up walnuts. He blamed it on his bad knee, but I noticed over the years that he was better in the role of directing others – a bossy-boots. Nevertheless, when he was around, things got done, including picking up walnuts. Under his direction, we picked up dozens of bags of walnuts.

Our hands would be dyed a distinct yellow-brown from picking up the walnuts. Not only ours. Many of the kids in school were just like us and, our all hands would stay dyed for several days. You cannot simply wash walnut stain away -- it must wear off. The more walnuts you pick up, the browner your hands would get. The boys were proud of the deep color, except when they sometimes served as a basis for racial slurs. The girls were less proud of this look, and more likely to wear gloves while picking up walnuts to keep the dye to a minimum.

Papa Gill, seeing my stained hands one day, told me that there was once a market for dye extracted from walnut skins. The walnut juice was the basis for the dyes used for giving khaki britches their distinct brown color.

Our walnuts weren't headed for the dye market, however. We'd take them to a feed store in Lewisburg that bought walnuts. We'd back the truck up to the loading dock and unload the walnuts, a bag at a time, and a worker would feed them into a walnut dehuller which made a loud grinding noise and would spit the green-black hulls out on one side and the denuded walnuts would roll out on the other side.

The newly dehulled walnuts were black, except they still had some of the black-greenish hull tissue clinging to the hard, rough nut. It took about three bags of walnuts-with-hulls to make one bag of dehulled walnuts. The bags of walnuts were weighed, and our payment was calculated. A day's work, including the ride to Lewisburg, would net us nine or ten dollars. It seemed like a lot of work for a few bucks, but it seemed to impress the Wilsons.

Some of the walnuts were saved back for us to eat. Since we didn't have a dehuller, we had to figure some other way to get the hulls off. I remember Papa Jepp's method was to loosely fill a burlap sack and take his tobacco mallet and bash on it until the hulls were mashed off and the nuts freed from their jackets. We sometimes ran them through a hand-operated corn sheller which did only a partial job and gunked up the machine.

My favorite method for dehulling walnuts is to spread them in the driveway and let routine traffic dehull them. The walnut is hard enough that it could withstand the pressure of tires without breaking. After a few days of being run over, we'd go out to the driveway and pick up the nuts. Usually we'd eat a few right then, but the same hardness that protected them from tire damage also made them difficult to crack. The best method is to put them in a vice, but we most often put them on a flat rock and bashed them with another rock, or a hammer if we had one. The bashing method usually ruined a little of the meat, but you could pick most of it out.

Fresh walnut has a unique flavor. It is surprisingly strong, but a little buttery and a little wild. It is so much trouble to get out, thar I'll go for years without eating any, then I'll try one and want more. It becomes an interesting challenge to wrestle enough meat out of several nuts to satisfy the hunger

As the fall season progresses, more and more of the walnuts that you find have blackened skins and are already partially dehulled. The only reason they are still around is that the squirrels have rejected them. If you bash one open, it is empty. The good ones are buried or hidden in hollow trees or under rocks. If you find one after a squirrel has extracted the meat it will give you a whole new respect for the ability of these animals to use their hands and strong teeth and probing tongue to wrestle sustenance from an almost impossible source.

Remains of Sawmill that Daddy used to make lumber from the trees on our farm.

I've discovered a company, Hammons, that sells all the black walnuts you want or need. Much easier than fooling with them yourself.

Hickory and Chestnuts

Hickory nuts and Chestnuts were the other two major nuts that were to be found in the forests of the Cane Creek valley. Hickory nuts come large and small, with the large ones being easier to dehull and crack, but their flesh isn't as tasty. Hickory trees grow large and straight and have distinctive shaggy bark. They make good baseball bats and hoe handles, but hickory is absolutely the worst wood from which to make fence posts. The farm I bought in '82 had been cross fenced with hickory posts. Within two years, every one of the posts was either totally or completely rotten. I had to either take the fences down or replace the posts with metal or Bodock.

Chestnuts are a sad story. In the early part of the twentieth century, one of the most abundant trees in the forests of the Cane Creek Valley were Chestnuts. Want to guess where the Chestnut Ridge got its name? They were huge, beautiful trees. I've seen stumps that were five feet in diameter. They were almost completely wiped out in the early part of the century by Chestnut blight. The few that were left were young and scrawny, and most eventually got the disease and died themselves.

Chinese Chestnuts have replaced them in some places and the foresters still find enough old stock Chestnuts to keep from having to say they are extinct. They might as well be for all the good they are doing along the Cane. In recent years, scientists have been breeding a hardy variety of Chestnut that may someday lead to a resurgence of this forest giant.

Bodock

The Arbor Day Foundation recently had a contest to determine the most popular tree in America. Oak won, but I would have voted for Bodock.

This is a tree of many names. Bois d'arc is the real name, but there's not enough French in the middle Tennessean's dialect to allow that name to exist without being corrupted into Bodock. The original offers one advantage in that it serves to remind that the wood was prized by Indians as material for making bows.

Bodock tree showing wide canopy with very few straight branches.

Osage-orange is the name widely used in polite company, and the one that you need to use if you look it up. The scientific name is *Maclura pomifer, (Rafinesque) Schneider.*

The other common name is Hedge Apple, which is also a clue as to its history. It was native to eastern Texas, Arkansas, and west Tennessee, but its range got dramatically expanded when someone got the bright idea that its fast-growing nature and a natural tendency to form a dense thicket would allow it to make a living fence, a hedge. This would possibly be a workable idea if Cane Creek farming were like English farming, where a farmer

Hedge Apple – the fruit of the Bodock tree

could allot time to tend to a living hedgerow. This isn't in the nature of the Cane Creek farmer and never will be. That's just as well; I doubt if the Bodock is naturally adapted to being suppressed to the degree that it would take to make a manageable hedgerow. The nature of the Bodock is more to spread over more and more land.

The spreading is true in at least two ways. First, individual trees grow large but not particularly tall. Instead they send their branches out over an impressive area – spreading their branches. It isn't unusual to see a tree that covers three or four hundred square feet. They live for a long time and continue to spread more with each additional year.

The other way they spread is by those so-called apples. The fruit of the Bodock are green, about the size of a grapefruit. The covering of the apples is distinctly bumpy in such a way as to invite comparison with a brain. One book describes the fruit as a 'large green ball, turning yellow when ripe, juice milky, frequently causing skin rash on exposure. There are 30,800 clean seeds per kilogram.' It makes me wonder who, how or why anyone would count the number of seeds produced by a Bodock.

Another source says 'The coarse fibrous texture and sticky, bitter, milky juice makes these fruits unpalatable to man or beast.' This is probably true as far as man and most beasts, but squirrels, particularly red squirrels, like eating the seeds. It is not at all unusual in the Fall to come upon a big red squirrel busily tearing the tough, fibrous rind to get at the seeds. They like to do this while perched on a large stump or rock, presumably so they can watch for predators while they eat. It is likely that squirrels bury whatever seeds they can't eat at a given time, adding considerably to the spread of the Bodock.

Squirrel hunters look for the piles of Hedge Apple leavings when searching for squirrels to shoot for frying purposes.

The fruit has one redeeming characteristic - roaches hate it! This has been known as an 'old wives' tale' for a long time, but entomologists at Iowa State have given us scientific proof. They compared it to chemical repellents and determined that 'The repellent actions of Osage-orange and Catnip have been confirmed.' Their study also looked at the effects on mosquitoes; I think it might be interesting if we could stop mosquitoes by simply rubbing ourselves down with Bodock juice. It might even be better if we could stop mosquito growth by heaving a few Bodock apples into the watery breeding places of mosquitoes.

I've heard some farmers put Bodock apples in farm equipment to keep mice from nesting and chewing up wire coverings.

Another old wives' tale was related by Mr. Wilson every Fall when the subject of hedge apples would come up. He said he'd always heard that, if a cow ate a hedge apple, she'd go dry. I thought then (and now) that a cow would have to be pretty danged hungry before she'd consider eating a hedge apple. If she was that hungry, she'd probably be near starvation ,and starving cows give very little milk.

The wood of the Bodock is a thing of wonder. It is a beautiful shade of nut-yellow when it is cut, but quickly turns into a rich greenish brown. It makes beautiful furniture but is not widely used because it is so hard that it is difficult to work, dulls saws quickly, and has a tendency to split.

Bodock wood, being dense, is full of energy and makes good firewood for a furnace, but it is not good for fireplaces because it frequently pops, sometimes throwing fiery cinders onto the hearth.

Posts made from Bodock last for decades because they have a remarkable ability to avoid rot. Most kinds of wood deteriorate with constant contact with soil. Bodock seems to last even better beneath the soil. The only real problem with Bodock posts is that they are usually crooked. The only time they grow straight is in a thicket. The more open the area they grow in, the more likely they will have a crooked, spreading growth pattern. Most of the posts, however, have a side that is straight enough to tack barbed wire on.

When you build a fence of green Bodock posts in the Winter, the posts will demonstrate the notable will of Bodock to live by putting out branches and leaves the following Spring. Once, Papa Jepp used Bodock posts to build a fence along the road. The posts put out enough branches and leaves that it looked like he was putting out a hedgerow instead of a fence.

It is easy to drive nails and fence staples into green Bodock, but it is practically impossible to drive a fence staple into cured Bodock. This makes it difficult to repair fences built of Bodock if the repair involves driving long staples.

If the spread of Bodock trees is not being constantly controlled by chemicals and bush-hogging, they will take over a farm. If they take over, it is hard to take the land back. Probably the best method is to use a bulldozer. Luckily, Bodock trees have a relatively small root system, and it is not difficult for a good dozer operator to turn acres of Bodock into a pile of burnable brush.

When I first bought my farm, I didn't have enough money to pay for a bulldozer to clean land. My choices were to leave the Bodock or cut the trees with a chainsaw. The person who elects to cut Bodock with a chainsaw is in for an interesting and lengthy experience. First, there is apparel. Long sleeves and leather gloves are highly recommended because the Bodock has a characteristic I have failed to mention to this point: thorns. The thorns are small but insidious. They are the worst on young growth and find ways to whip the person that is trying to cut them. They penetrate clothes to make painful scratches on legs, hands, and face. It stings the worst when one of the thorns sticks into a cheek or an ear.

Sometimes the tip of the thorns breaks off under the skin. They are small enough that they are exceedingly difficult to remove, and they have enough of some kind of toxic resin or oil to make them painful and slow to heal.

Bodocks grow at odd angles, often with multiple trunks and unpredictable twists, making it difficult to predict which way the tree will fall. This makes Bodocks tricky to cut because the tree may lean the wrong way as it begins to fall and may pinch the saw into immobility. Sometimes, if the saw handler is quick, he can wrest it free before it is trapped, but if it is a big tree, the saw can be so stuck that the only way to get it out is to work it out with wedges or use another saw.

The chain saw strategy includes cutting the tree close to the ground, leaving as little stump as possible. The stump will last for decades and leaving more than a few inches will result in a seasoned piece of wood, as hard as any wood that exists, and well-anchored by roots that almost never rot. This stump will be a constant irritant, exactly the right size to be hit by a bush-hog blade, which makes a terrific noise and can even break the blade or loosen its mountings, eventually resulting in a difficult repair job.

The stumps are usually too low to bulldoze away, so the only way to get rid of them is to saw them closer to the ground. Chain sawing a seasoned Bodock stump presents a unique challenge. The wood is so hard that it is common to see sparks fly, so it is no surprise that blades are quickly dulled when cutting seasoned Bodock. Also, since the goal is to get close to the ground, it is difficult to avoid getting the chain saw blade into dirt and nothing dulls a blade faster than sawing dirt (except sawing rocks or steel wire).

Bodock has a phenomenal will to live. Fresh stumps must be liberally painted with a combination of used crank case oil and tree poison. If poison isn't applied the stump will sprout with new growth the next spring. It isn't unusual for a six-inch trunk to have eight to twelve sprouts that shoot up and quickly turn into woody whips of thorny sapling. Within a year, these sprouts may be twelve to fifteen feet tall. In another year or two, the stump will be the base for a mini-forest of trees, growing upwards and outwards as it struggles to replenish a claim over its corner of the world.

Maybe the fact that the tree is known by so many names is a clue to its adaptable nature. It grows fast and strong and will compete with walnuts and oaks to dominate good land, but it thrives almost as well on shallow, rocky soil, taking on Cedars and Honey Locusts in their challenging habitat.

The Bodock is many things, but, most of all, it is beautiful. Most trees have straight trunks that shoot toward the sky as if trying to break their bond with the earth. The Bodock appears to flow from the earth with grace and sinewy strength. The leaves, slick and deep green, are a rich crown for the sinewy trunk. These same leaves turn bright yellow in late October, offering a bright moment before the tree's winter rest. The apples are unobtrusive until they fall, and then are so interesting that it serves little to note their lack of beauty.

Cedars

Cedars are the dominant evergreen of the Cane Creek Valley. Technically, they are a variety of Juniper, but people in Tennessee will look at you funny if you refer to these ubiquitous trees as anything other than Cedars.

I once invited Tom Hall, the District Forester, and Stanley Lyons, the Extension Agent, to advise me about managing the woods on my farm. Tom's advice was to do nothing except let the Cedars grow and harvest them when they got large enough. I'm sure he was correct and was trying to be helpful, but the hidden message was, "Face it, Warren, your land is too poor to grow anything but Cedars."

Cedars grow practically anywhere, but they prosper on the rocky hills along the Cane Creek and its tributaries. They grow slow and strong, with deep green foliage and a straight trunk. Cedar posts are second only to Bodock in their ability to last and they are straighter and easier to work with. One caveat: young Cedar with mostly white wood will not last. It is the older Cedars with large cross-sections of red wood that make the long-lasting posts. The red wood is not particularly hard, but it is loaded with a resin that is sticky when the Cedars are cut but dries in the cured wood and renders the wood unattractive to the creatures which cause rot.

Old, tall Cedars make excellent poles for building barns. It was not difficult at all to find twenty and even thirty-foot poles that could be set in the soil three or four feet deep and still leave enough room for building a barn tall enough to hang three tiers of tobacco or build a loft to hold tons of hay.

The smaller Cedar poles, with plenty of red wood, but not barn pole size, make beautiful, long-lasting fences. These were common before wire fences came along. The zig-zag variety was the most common, probably because of the inherent stability of the triangular construction, but they could be built straight if double posts are placed to hold the poles in place. Either way, the poles are interlaced at the ends, with big ends and little ends carefully balanced to keep the fence level. Black Locusts work almost as well as Cedars for pole fences, some might think better, because they split easily into straight-sided poles which stack straighter than Cedars. Daddy was great at splitting posts. When he swung the axe, the blade hit in exactly the right place to make uniformly sized posts. Same for firewood.

Many of the fences and barns in the Cane Creek Valley were once built of Cedar but these disappeared when the pencil industry came to Lewisburg and Shelbyville. Shelbyville is known as the Pencil Capitol of the World, but the plants in Lewisburg made as many or more pencils. They make the plain old yellow #2, the thick carpenter pencils, and they make more make-up and eyebrow pencils than can be imagined.

Papa Gill told me that, when the pencil factories started, they first took all the living Cedars that were big enough to have red wood. Then, when the Cedar trees were exhausted, the wood buyers started buying fences and barns. The deal they'd offer for the fences was to replace the fences with new woven wire and a barbed wire along the top. This was a great deal for the landowner. Cedar fences are attractive, but they are difficult to maintain, don't hold livestock very well, and are hard to mow around.

After the pencil factories took all the fences and barns, they were desperate for wood. Somebody discovered California Incense Cedar and started shipping it in by the boat load. The industry, which had originally located in lower middle Tennessee because of the Cedars, was able to stay despite having to bring wood from California.

There is a market for Cedars to be turned into furniture and closet lining as well as a very small market every year in Shelbyville for people who bring freshly sharpened knives to the Celebration so they can whittle Cedar sticks into little piles of curly shavings.

Warren and Papa Gill

On our bookshelf is a picture of me with Papa Gill when I was about three years old. He has a knife in his right hand and a short Cedar stick in his left hand. He was showing me how to whittle, clearly enjoying himself, and I was absorbed with what he was doing. I remember very well how totally fascinating I used to find whittling. I could watch either Papa Jepp or Papa Gill whittle for long stretches of time and listen to the many things they'd find to talk about.

The place where one could most likely find both Papa Jepp and Papa Gill whittling at the same time was on the corner of the square, under an old Silver Maple in front of the local Mortuary. There would often be four or five old men, maybe more, and they'd be sitting in chairs provided by Mr. Russell Beasley who was the proprietor of the funeral parlor. I think it may have been

Aunt Mary Neil who first referred to the group of whittlers as the 'Mortuary Club.'

Many times, when I was little, Papa Jepp would do his farm work until the middle of the afternoon, and then he'd stop by the house and pick up Alan and me (Gloria would join in later) and take us to town. He'd often have to go by the feed store first and let Paul Barham sell him some scratch feed for the chickens or something. Then we'd go to the square, and he'd sit down to visit with the other members of the Mortuary Club. If Papa Gill wasn't already there, he'd often show up in a few minutes.

All the old gentlemen would be dressed in plain clothes, except Papa Gill. He always wore his black suit and a tie.

As Papa Jepp approached the group, he'd slip his hand into his back pocket and smoothly pull out his knife. He always used a Case for whittling, but he often had at least one other knife in his pocket that no one ever saw. He claimed the extra knife was for trading. Every once in a while, according to his story, somebody would come along and offer to trade the knife in their pocket for the knife in his pocket. Such an offer didn't come along often, but when it did, he was ready. He'd agree to the deal and pull out a knife he'd probably bought for a quarter. I remember asking him how often that kind of deal came along. He didn't like the question but told me it had happened at least once and gave me the name of the person who'd made the offer. For some reason, it struck me that carrying an extra knife for a lifetime in order to beat one person out of a knife was bit extreme, but in retrospect I'm not so certain. After all, he was able to tell that story probably a couple of hundred times.

As Papa Jepp approached, Mr. Russ would walk into the Funeral Parlor and grab another old wooden folding chair and unfold it as he said, "This one do all right, Jepp?"

"It'll do fine." And before another moment passed, Papa Jepp would be comfortably seated and his knife would be smoothly shaving aromatic curls from the Cedar stick. He always had several sticks in his truck and others in various places around the house and tool shed and never took a chance that a whittling opportunity would occur without him having a stick.

Papa Jepp whittled fast and didn't pay much attention to the result. Papa Gill treated whittling more seriously. Both men spent time looking for good Cedar, but Papa Gill was more demanding. He wanted a tight, straight grain with no knots or imperfections. He cut his sticks shorter and whittled them more carefully, making small, thin curls. Each curl was practically identical to the last. His knife was always razor sharp, and his stick that was usually almost perfectly round.

If you asked any of the whittlers what they were making, the answer was likely to be, "a mess." Papa Jepp never made anything but a mess, but, occasionally, Papa Gill would carve something, usually it was simply a little tree, which he made by allowing the curls to accumulate on one end of the stick. Mama Gill stuck these little Cedar curl trees in pots with house plants.

*****

"Christmas ain't Christmas without a Cedar tree." I'm not sure if this quote can be attributed because it, or something like it, has probably been said by too many people, too many times, and in too many places for accurate attribution. Cedar trees aren't as pretty as store-bought trees, or as shapely as the white pines so carefully tended and trimmed by tree farmers, but Cedars have two distinct advantages: they are cheap and they smell like Christmas.

The cheap part is easy to explain. It is simple supply and demand. Anything as plentiful as Cedar trees cannot possibly be made into something expensive. People just aren't that dumb. Cedars are also a little drab. They are an evergreen, so they stay green all the time, but it isn't a very bright green, especially around Christmas time.

Cedars can be described as a drab, which makes it all the more remarkable that they look so pretty after they are decorated.

The female trees have an abundance of small grey-green berries; these don't add much to their beauty but they do increase the intensity of their pleasant Cedar smell.

The smell is the important part. Scientists say the roots of the sense of smell go deep in the brain so an old familiar smell can revive lost memories. The link between the Cedar smell and Christmas has been forged over many generations in the Cane Creek Valley, which probably explains why the smell of Cedar sometimes lets me see Granny like she was when I was a child, busily cooking Christmas dinner, directing where the presents were put and who was to take out the trash. Papa Jepp may have ruled the farm, but Granny ruled the house, and this was never clearer than during Christmas.

Granny wanted her Cedar Christmas tree to be very wide and if Papa Jepp failed to get one 'fat' enough, he'd be heading back to find another. Mama was just as particular about her tree but didn't want it 'fat.' Instead, she wanted one tall and thin, with thick foliage and the right size to fit in front of the east-facing window in the living room. The ceilings in that room are almost fourteen feet tall. I encourage you to go out and try to find a perfectly shaped fourteen-foot Cedar. Hint: take a sack lunch.

Finding the perfect Cedar Christmas tree is not an easy task. There may be millions of Cedar trees, but not many of them come with a specific, ideal shape. A tree that looks good from the road is often flawed upon close inspection. For every good tree, at least a dozen others have a double trunk (making them impossible to cut without obtaining unusable part-trees) or are too small or too big or simply ugly.

It was always an adventure to go Christmas tree hunting, but my most memorable Christmas tree hunt wasn't the most successful. I was twelve and Alan was ten. Daddy was working long hours at the arsenal and hadn't had time to look for a tree, so I decided that Alan and I could do it. I'd been given a hatchet for Christmas the year before, so I figured I had all I needed to get the job done.

Alan didn't seem quite as confident that this was a good idea. He knew that some of my ideas didn't always go as planned, but the prospect of whacking a tree trunk with the hatchet finally won him over. It was cold, and there had been intermittent rain with some mixed-in snow, so Mama made us put on our heaviest coats and wrap our necks with scarves. Caps and gloves finished our ensembles. We trudged up the hill and walked and looked until it was almost dark. We had to find something soon or go home empty handed.

Finally, we found it. It was a little further from home than we would have preferred because the only way we had to move the tree was to drag it. Never mind; the tree was beautiful. It was exactly the right shape and possibly the right height, although we weren't too expert at estimating height. I took the first turn at chopping the tree. The hatchet wasn't very large and took only a small bite with each swing, but there was progress. I took about ten strokes, then turned it over to Alan. He whacked it for a while, opening the somewhat irregular notch to about an inch. His strokes were better aimed than mine, but neither of us had developed much technique at that time.

The tree was larger than we thought. It took us turn after turn. We'd hit one side for a while, then the other. Our heavy coats impeded our ability to swing but we dared not take them off for fear that we'd catch cold and be sick at Christmas time. I began to think my hatchet wasn't as good as I'd hoped. Alan was getting tired of the game and suggested we go back and bring Daddy up here on Saturday. I wasn't about to give up that easily.

I attacked the tree with renewed vigor. To my surprise, the tree started to lean. A few more chops and the tree fell to the ground.

It took some more chopping to completely sever the tree, but it finally came loose, and we were ready to start dragging.

With both of us pulling with all our strength, the tree moved. It didn't move fast, but it moved. We'd pull a few feet, then we would need to stop for a minute to see if we could think of an easier way to get this job done. We couldn't, so we'd pull again.

Then we came to the fence. It was a simple barbed wire fence, pulled tight and nailed to Hackberry and Bodock trees. We'd previously climbed it almost without thinking, but now it presented an almost insurmountable obstacle. I tried to think of some way around it, but I knew the nearest gate was almost a half mile away and we'd never be able to pull the tree that far.

Alan suggested waiting for Daddy. That made me determined to get the job done so I pulled the tree up to the fence, and with Alan's help, rested the trunk on the top wire. Then I crossed over and started to pull it over. I made a little progress, but not much. After another attempt or two, I asked Alan to lift and push. I'm not exactly sure what he did, but it worked. With him pushing upward on the tree and me pulling, we got it over!

I felt like we'd won a major battle, because the rest of the pull was downhill. We jumped to it and made great progress for about a hundred yards. We stopped to rest for a moment and turned to admire our tree. We both agreed that it looked better now that it was out in the open. The house was only a short pull away, but we were tired. We decided we'd done enough. The tree could stay right there, and we'd bring Daddy out to see it the next day and he could help us with the last pull.

We told him all about it when he came home after work. We made it sound like it had been easy, but still let it show that we were proud of the job we'd done. The next morning, Daddy had to go to work for a while, but promised he'd be home before dark, so we bided our time, not worrying about the tree. We'd take care of it when Daddy got home.

When Daddy got home that afternoon, the weather had gotten better, but it was still cold. We threw on our coats and led him to where we'd left the tree. As we drew close, I could see something was wrong. It had lost its color. It still had the same perfect shape but was no longer green. My first thought was that it had faded for some unknown reason, but as we got closer, I could tell that all the green had disappeared. Daddy quickly solved the mystery. "The cows ate your tree, boys."

I wanted to deny it. We should have pulled it into the yard. It would've only been another fifty yards. It just never occurred to me that a cow would eat the green from a Cedar tree. I'd never seen them eat the Cedar trees that grew in the pasture. I said something like, "Cows don't eat Cedars."

He said, "Well, they ate this one. Looks like it was a nice one." The cows had simply nipped the green parts and left all the woody parts, even the small branches and twigs. The shape of the tree form was unchanged - it simply had no green foliage. It even occurred to me that we could still use it by simply painting it green but some foolishness filter in my brain thankfully prevented me from saying that out loud.

Daddy said, "Don't worry about. We'll get another one." We did just that. With Daddy helping we had another tree in the back of the truck under Mama's critical eye in less than hour and a half. This tree passed with flying colors.

The skeletal cow-eaten tree stayed in the pasture all winter. I saw it many times. At first, it caused a twang of pain to see it. It seemed wasteful to take a tree, then let it fail to meet its purpose. The poor tree never got to be decorated. It never got to preside over a house full of happy people exchanging gifts. Then I'd remember it was just a tree. I'd also remember that last fifty yards.

Cousins

The Cane Creek would run whether people lived in its valley or not. It ran long before people came, and it will continue to run when the people are gone. Generations of young people have incidentally learned to swim as they played in the Cane Creek swimming holes. Sometimes people have dirtied the creek, but more often they have done their best to keep it clean. People have forded it, bridged it, torn its banks, and robbed the gravel from its shoals, but they have mostly built their lives around it as they work and play.

Family is the foundation upon which lives are built. The people of the Cane Creek valley understand the value of family and not only immediate family. Extended family gives stability to local social structure. Many of the Sunday afternoon porch conversations were about who was kin to whom, and all you had to do was listen for a while and you'd understand that a lot of people up and down the Cane Creek valley were kin to one another or kin to someone else who was kin to another someone.

After Lissa and I were married, Papa Jepp told us how we are blood kin through the Conaways or Cranes or Reeves or some such. It was a long, convoluted pathway, and I don't pretend to remember it. The point is, it would be hard for two people whose families had been in Lincoln County for a century or two to not have some blood in common. Such is the nature of kinship.

It is also the nature of kinship that kin helps kin. If someone opens a store, kinfolk tend to trade there. If someone's house burns, kinfolk help find another place and give furniture to help the poor cousin get another start. If someone starts selling insurance, guess what? Sometimes helping kin is a headache.

Uncle Edward stayed home and farmed during the war. Farmers were needed for the war effort as much as soldiers and he was a strong and good farmer. He and Aunt Margaret started working the farm in those difficult years and they also started our generation of cousins when Gail was born. Gail, being the oldest, was often our guardian and protector. She was the wise cousin that we could trust to give the best advice because she'd already learned the best way. We always knew she was smart, but I found out later that many of the teachers in Petersburg considered her one of the most intelligent people they ever taught.

After Gail came Eddie. Eddie was, in many ways, my earliest role model. For one thing, he was amazingly strong. I remember him picking up a horseshoe in Papa Gill's barn lot and bending it. It was old and worn thin, but bending it was still impressive. Eddie was 'cool.' He was the first person that ever let me drive a tractor while I sat on his lap. One day, when he was supposed to be mowing our lawn, he took a break and built us a tree house that we played in for years. For a cousin who was five years older, he never seemed to be impatient with his younger cousins. He seemed to enjoy our company. He'd take me rabbit hunting and patiently teach me how to refine my shooting technique. Once, he was charged to baby sit me and he told me to get in his car and we'd go for a drive. After we got away from adult eyes, he let me drive up the dirt road to his best friend, George Warren's, house. George was another cousin, but not a first cousin. This was my first time to drive a car and I was thrilled, but that wasn't the only excitement of that day. When we got to George's house, George walked out to the car and Eddie took over the driver's seat. He and

George were chatting about the cool things that teenagers talked about, and even tried to keep me in the conversation, although I wasn't in their league of cool. The coolest part of all was when they pulled over to the side of the road and pulled out cigarettes. They were smoking! Eddie even offered to let me have a drag. I declined and felt bad about it, but neither of them acted as if it mattered one way or other. I could do what I wanted because I was cool enough to hang out with them. It all sounds hokey now, but it was deeply important to me at the time.

Peggy was born on December 3, 1951, which made her less than a year younger than me. That proximity in age made us good buddies. When we got together, we were always coming up with ideas for exploring something, whether it was Mama Gill's basement, the cow barn, or an old tenant house.

One day, Uncle Edward, Eddie, and I went rabbit hunting in the field behind Papa Gill's house. They had shotguns, and I only had an old twenty-two rifle, but I was thrilled to get to go hunting. We'd been out for a while without seeing anything, when I looked to my right and saw a rabbit hiding in some sage grass. I quickly aimed and shot it in the head. Uncle Edward and Eddie both made a big deal out of my good shot. Naturally, I was practically giddy with pride and I couldn't wait to tell Peggy. When we got back to the house, we started cleaning my rabbit and another one that Eddie shot, but Eddie's rabbit had slightly revived, so I picked it up and whacked its head against a big Ash tree to 'put in out of its misery.' Just then Peggy came around the corner of the house and saw me killing the rabbit. She was at first shocked, then started crying and ran away. I felt terrible, not about killing a rabbit, but for making Peggy so unhappy.

We finished skinning and cleaning the rabbits when I had a bright idea for making up with Peggy. Everybody knows that a rabbit's foot is good luck, so I cut off one of the rabbit's front paws and hid it in my hand. When I saw Peggy under an old Maple, she was no longer crying. I was relieved and almost ran over to her. She smiled and was about to say something when I held out my hand to give her the fresh rabbit's foot. It took her a second to realize what I was holding. Stupid! I cringe even now thinking about how she started crying and ran away again. Stupid! Stupid! Stupid!

Some of our best adventures were in the cemetery. In our early childhood, Uncle Edward and Aunt Margaret's house was directly across the street from the Old Orchard Cemetery which is the main burial place for the citizens of Petersburg and the Cane Creek valley.

A cemetery may be a place of poignance or sadness or even celebration of rich lives. It may draw curious descendants of long dead forebears or visitors seeking information about extended family. Many of our family are buried there. There are several Gills, but there are also many Warrens, Bledsoes, Moores, Rhodes and other families that are our kin. We knew that we had family under the ground in that field but, in those years, to those of us who were too young to have been touched by death, the cemetery was simply a grand playground.

I'm not sure Alan liked playing in the cemetery as much as the rest of us, possibly because of the ghost stories that Gail and Eddie used to tell. None of us liked to admit how much the stories scared us. The fact that Gail and Eddie managed to set most of the stories within the Old Orchard Cemetery made them even more frightening. Alan was as tough as or tougher than any of us, but he was also younger, and the look I sometimes saw on his face made me think the stories frightened him more. It seems somehow appropriate that Alan has become a member of the committee that manages the Old Orchard Cemetery. To this day, he has to go to Decoration Day (Mother's Day) for a service and to collect money from visitors for upkeep of the cemetery.

Then, John and Jane, the twins, came along. The fact that they were twins was amazing enough, but their development into little people became a factor in our play. As we planned our adventures, Peggy, Alan and I had to take them into account, because they were there. When Gloria became big enough to start toddling after us, it added yet another complication. Sometimes our adventures were as simple as walking around until the little ones got tired of us or simply figured out something more fun than following the big kids.

*****

The creek pulls people. Many and varied are the reasons for visiting the creek, but it is often as simple as 'let's go to the creek.' Mama and Aunt Mary Neil were always finding ways to get our families together. There was a distinct contrast in the nature of the town visits and the country visits. Naturally, it was an adventure for us to go to the city of Huntsville. They had stores and restaurants, especially my first McDonald's, and things to see and do, but mostly they had a swimming pool. The swimming pool was a study in cleanliness and regularity. There were no sharp-edged rocks that could cut tender feet and stepping in a fresh cow flop was out of the question at the swimming pool. The concrete was smooth and the water clear and carefully filtered to make sure it stayed that way. We even had to take a shower before we got in the pool (which I thought was crazy!). The chlorinated water made our eyes red, but it was a small price to pay for the extraordinary cleanliness.

I personally didn't think the city pool was anywhere near as good as the Cane Creek. The pool didn't even have frogs, snakes or fish! And you didn't have to shower before you went in, but, for some unknown reason, Mama made us take a bath after we'd been in the creek.

Mostly, we loved visiting the wonderful city of Huntsville. It was great, except the milk. I felt sorry for my city cousins because they had to drink store-bought milk. Also, the houses were a little crowded together.

It was a nice coincidence that Aunt Mary Neil and Mama managed to have daughters of about the same age. Gloria and Carol were close from the time they were babies and getting these two together was often at least part of the excuse for visits. The fact that Gloria had two older brothers, Alan and me, and Carol had two younger brothers, Bob and Bill, only made the group of us appear like a large continuum of a family when we gathered, and, when we gathered, the kids gravitated toward water.

When Uncle Jim and Aunt Betty added Rock and Julie to the family mix, we had two more reasons to visit the creek. No kid can resist the creek. It is impossible. Quiet, introspective kids may be attracted for different reasons than rambunctious, mischievous kids, but all find something to like.

There was once a fad toy called a Slip 'n Slide that was a long sheet of plastic with a hose fitting at one end that allowed water to stream out creating a long slippery surface for kids to run and slide. Could anything be better for a kid? -- The answer is an emphatic 'yes!' A creek is clearly better than a Slip 'n Slide. Creeks last longer and are naturally more slippery than plastic. There is more sliding surface and they don't run up your parent's water bill. I remember thinking how sad the lives of people must be who think that a Slip 'n Slide is some kind of neat toy.

There was one place in the creek where exactly the right combination of flat rock bottom, shallow water, and smooth rock bank made a place that was better than a dozen Slip 'n Slides put together. When conditions were right, the creek bottom became slicker than a combination of WD-40 and thirty weight oil. I think the correct expression is 'slicker 'n snot.' A kid could back eight or ten feet up the bank and get up as much speed as their legs and spirit could muster, then hit the sliding place and ski for at least a dozen feet.

Of course, at least half of the efforts resulted in disaster. If churning feet snagged a dry spot, the kid somersaulted into the water, usually with a twist that kept their head from getting bashed. Most common was hitting the water at a funky angle with a quick drop to the rear. The subsequent bottom slide was still thrilling but not as long.

A favorite of the young kids was to get an older kid to swing them into the creek at the sliding place. Gloria was especially clever at getting me to throw her and Carol or whatever little girls were around into the creek Slip 'n Slide. Several times, I thought they were going to break my back (or ear drums) before I was able to convince the little squealers that I had to go do something else.

*****

When any cousins or friends came to the country, the creek drew us like a magnet. It may not be as clean as a city swimming pool, but it offered its own delights. For one thing, you didn't have to swim or even pretend to want to swim. A simple walk along the bank with a few cousins or friends is a great way to catch up on news and gossip.

When the cousins or friends would show up, they'd often be wearing their good clothes. There was usually a reason. Maybe they'd come straight from church, or it might be Easter or the Fourth of July. Maybe they simply didn't think about putting on old clothes. Having on good clothes never prevented us from wanting to go to the creek. Many a trip to the creek started with a promise that 'We won't get our clothes wet.' Mama understood that a promise like that was made to be broken, and the mother of the visitor would usually give in.

We'd go to the creek and start throwing rocks and looking for snakes. Then somebody would decide a little wading wouldn't hurt a thing and start taking off shoes and socks and rolling up pants.

By the time the first kid was getting his toes wet, the rest were shucking their footwear. In a moment, we'd all be gingerly wading the shallow pools. These pools were often very slick-bottomed (see earlier snotty description) and kids can't help but have a little horse play. This combination inevitably led to some kid falling into the creek. That kid very rarely wanted to face punishment alone and would do his or her best to make certain someone else also fell. Before it was over, every kid on the creek would be soaked.

The planned swimming parties were maybe the best. Somebody would look at the weather and decide it would be a good day to visit the Gills. Phone calls would be made, and some combination of cousins and friends would show up with swimming suits and towels and swimming goggles and all the other stuff that makes for fun at the creek

Conditions weren't always perfect. In the summer, during dog days, the water sometimes didn't flow much and turned scummy and dank -- not exactly good for swimming. Sometimes, after a rain, there would be too much water for safe swimming, but on many days Cane Creek was and is the most perfect place in the world.

Sometimes the creek got almost dry. Then it was fun to look for rocks and odd pieces of wood or discarded metal. One time, it got so dry that the well at the house quick working, and Daddy had to haul in drinking / cooking / flushing water. To save water, he'd grab a bar of soap and a few raggedy towels and take the kids to the creek for our bath. Remember, if the well was dry, the creek water was also low, but there was one pool down from the house that was spring fed and always had some water. Since it was spring fed, it wasn't too nasty (not overly clean either). We'd all get naked and clean ourselves as best we could. I don't know if we got clean, but we had fun. Gloria was barely walking, so thankfully can't remember our nekkidness.

A group of kids in the creek is many things. It is sometimes loud, with girl shrieks as some boy splashes them or finds a snake that needs to be shown around. There is exploration as the kids traipse up the creek and down as far as their energy lasts and their courage holds up, as they leave familiar places and venture into strange lands owned by other farmers who may or may not welcome marauding groups of kids.

There were thrills. Vine swings are the classic source of creek thrills. Tarzan was a bad influence. For many years, every vine that grew near the creek was regarded as a potential swinging vine, and a hatchet was as important as shorts when a swimming trip was planned. The problem was that perfectly hanging vines were rarely found over those few pools of water that were deep enough to swing into. The solution was to hang a rope in the place where the vine ought to be. Many hours were spent in engineering these 'rope' vines, but the thrill of the swing over water and the final splashy flop made it well worthwhile.

Other thrills were more subtle. Sooner or later, someone would remember that our swimming hole was also a fishing hole. Where could all those fish be? There were plenty of fish hiding places along the creek banks and under big rocks. Why not just catch a few fish while exploring under the banks? Then, some boy would start sticking his arm under big rocks or into the deep holes under the bank. I say 'boy,' because I never saw a girl foolish enough to try such silliness.

Sometimes the holes under the bank were huge. If the back of the chamber under the bank couldn't be reached without diving, then the only thing to do was dive. Diving meant instantly leaving the world of light and air and entering a totally different world of darkness and water. There are catfish and suckers and sunfish and bream, but there are also snakes and snapping turtles. Most of the snakes are harmless, but the turtles are another story altogether.

Every country kid knows that snapping turtles can inflict a nasty bite.

Bringing up the subject of snapping turtles among country kids almost always gets tales going about catching a snapping turtle and teasing it until it bit a stick in two. The size of the stick is never exactly measured but is usually described as huge in the telling. To tell the truth, I tried it several times and never found a snapping turtle dumb enough to bite a stick. We were told such a fierce bite could easily remove a toe or finger. This is the kind of conversation that fires the imagination when you start putting hands with fingers under rocks and creek banks.

Snapping turtles are not an endangered species, or even rare. Some larger ponds may have dozens of them. They inhabit numerous locations along the creek bank. If turtles were as fiercely determined to snap fingers off as country boy gossip implied, one would reasonably expect to see an occasional missing finger or toe. Over the years, I've met many people who have lost fingers. I know of at least two people who have lost fingers while working cattle. Papa Jepp lost the ends of two fingers in a corn picker. Uncle Edward cut a finger off with a skilsaw and had it reattached (to his regret). Of all the missing fingers I have seen over the years, not even one was due to turtles. I have never met anyone who lost any part of any appendage to a snapping turtle.

People don't lose many digits to turtles because turtles don't want to waste time on fingers or toes. Turtles are naturally shy and reclusive, and if you invade their space, they do their best to escape. When they snap at a stick, it is because it is their only recourse after being prodded and poked. If fear of turtles is all that is preventing anyone from sticking their hand under a rock, I say go for it. Frankly, I'd like to meet someone with a turtle finger-ectomy.

Visiting

(Just to show we did social things besides play in the creek or go fishing)

We used to just go visiting. I don't think most people do such a thing as much anymore, but we used to do it all the time, especially Sunday afternoon. There might have been some reason, like a birthday or to deliver some item, but much of the time it was just to go for a visit.

The most common place to visit was to grandparents. Granny and Papa Jepp loved visits as much as anyone. Problem was, they were such energetic people, that they were more likely to go visiting than to wait around for people to visit them. If you planned to go visit them, you'd better call first or you'd likely find an empty house.

Papa and Mama Gill were a little more likely to be found at home on an average Sunday afternoon, and they were always proud to have us over. They lived in the old family home, a huge antebellum house with lots of places for hiding in, running through and, playing around. There were scary places, like parts of the basement or the upstairs front porch balcony (very high with rickety floorboards).

A typical visit to Mama and Papa Gills' could include anything from riding horses to exploring the farm to playing in hay to watching TV (they had the first one in the family). Their favorite show was Mitch Miller, then Lawrence Welk, which they thought was great music and which I thought was major tedium.

Mama Gill was diabetic, but still had more good sweets than anyone. Her pies were sweet, but more than that, they were robust in flavor. Even mincemeat, not my favorite in those days, was wonderful. Papa Gill would usually have a cache of "'Lion' peppermint sticks which were more to my taste.

Hattie Bell Porter was their maid. She looked like something like Aunt Jemima, and I thought she was wonderful. For one thing she could cook anything, probably using more lard in one meal than most people these days would eat in a year. I found out later that Hattie Bell enjoyed partying, and Papa Gill had to get her out of jail more than once.

During one visit, when I was a toddler, I was being a pest around Papa Gill. He was visiting with Uncle Allen and some other family men, and I began to bother him, so he reached down and gave me pinch. I started howling in pain. Uncle Allen, who loved telling this story, said I ran straight to Mama. She picked me up, consoled me, but also wondered what had happened. Since I'd come from where the men were visiting, she went in and asked. According to Uncle Allen, Papa Gill quite simply said with deep concern that he had no idea what was wrong with the poor child.

*****

A favorite place to visit was Uncle Hal and Aunt Sister's home. Aunt Sister's name was Margaret, but Papa Gill called her Sister so she became Aunt Sister (at least to me; Gloria and Alan don't remember it that way). She'd married Hal Moore many years before I knew them and both were quite old by the time I first knew them and both were always so gracious to visitors that they are always the first people who come to mind when I think of going visiting.

Uncle Hal was a small, bony thin man who usually sat in a rocking chair next to the hearth of their warm coal-burning fireplace and smoked a pipe as he gleefully joined in whatever conversation was going. Aunt Sister was also thin, with white hair, usually tightly bunned. She was always at pains to make everyone who visited, even children, as comfortable as they could be. I remember many things about those visits, but none well because I was so young, but the thing that probably impressed me most was the outdoor facility. By the mid-fifties, most people had installed indoor bathrooms, but not Uncle Hal. If you had to go, you had to take an outdoor stroll out the back door and down a well-worn path. It didn't matter if it was warm or ice cold, you had to 'do your business' outside in a wooden outhouse. I'm sure, if they'd wanted it, they could have had an indoor bathroom, but I think they'd gone that long in life with outdoor facilities, so they didn't want to change.

Uncle Hal and Aunt Sister's house was small, set on the downhill side of the Booneshill Road so it was almost hidden. The house was always warm and comfortable, neat and well-kept. There were beautiful flowers, and Uncle Hal kept a vegetable garden across the road. He always seemed to get more from his little garden patch than anyone would have thought possible.

Sometimes we'd visit Uncle Hal and Aunt Sister for no reason, but one occasion for visiting them was when Mary Gill came home. Mary Gill, their only child, was once married to someone who gave her the last name Kilgore and moved her to Iowa. She didn't stay with him long, but she had found a good job with Alcoa and had made friends up there, so she continued to live in Iowa where she still lives today. Her visits were a great joy then and continued to be a highlight of our lives for many years. She is a link with a rich past, and her zest for life makes her a joy to be around.

*****

Often, a visit to Uncle Hal and Aunt Sister was followed by a visit to Lucille and Clayton Scotts', which was just a mile or so on down Boonshill Road from Uncle Hal's. Miss Lucille was one of my favorite Sunday School teachers and so close to us that we called her Aunt Lucille for some years. I remember Mr. Clayton as an important person in the Petersburg First National Bank who sang in the Methodist Church choir with a beautiful tenor voice. He used to smoke but it started bothering him, so he quit. After a year or so, he started back, then got the cancer that killed him.

Miss Lucille lived on the farm where we visited her and drove a car until she was in her nineties and became almost totally blind, but she delighted visiting with anyone at any time.

*****

Visiting Belleville was a double-header. We'd visit Maw-Maw (Cora Crane - Granny Collier's mother) and Aunt Sue Wright (Papa Jepp's Step-Mother - he always called her Miss Sue). Granny and Papa Jepp both came from the Belleville area, and a trip to that community was always something of a walk down memory lane. They knew every inch of the little village and had a story to go with practically every house and barn.

Maw-Maw lived just a short walk from the Belleville store, in a small, white house that had a wire fence around it with a post next to the driveway that somebody said had been marked by hobos with a sign that indicated that the woman inside was good for a meal if you were willing to do a little work. She also had an old-fashioned phone, the kind with the ringer on the side. She used it as long as the phone company would let her and left it in place even after they'd forced her to get a new one.

Maw-Maw raised four daughters: Thelma (called Tutu), Bernice (called Aunt Bun), Carleen, and Neil (my Grandmother). These women were the liveliest set of older women I've ever seen, so I can't imagine what a hand-full they were as children. After Mam-Maw died, they had a squabble. The result was that Tutu didn't get along with Carleen and Bernice for a long time – until she died. Granny managed to get along with all of them.

Aunt Sue lived in a much larger house, but the old place was fairly run down. So was she by the time I knew her. In fact, as much as anybody I ever knew, she seemed to exemplify 'old.' She was deeply wrinkled with yellowed hair and swollen legs. She had a hard time getting around, but always knew where to find some toys for us to play with. I'll always associate Lincoln Logs with Aunt Sue. It seemed like I spent hours playing with them on her floor while the grown-ups talked on and on about nothing of any importance.

Back to the Creek
Uncle Jim gets snakebit

Lest someone accuse me of downplaying the role of snakes in causing creek problems, there is one story in which a snake caused a problem. When Uncle Jim was in his teens, he and Lyle Barham were always doing something together, and the Cane Creek was often involved. They ran trap lines and caught several mink and a bunch of muskrats. They often went fishing together and, on one spring night, they decided to go gigging. They had come by the house earlier to borrow one of Daddy's gigs and a flashlight, so we knew they were on the creek. They hadn't been gone long when they came back. Both the boys were out of breath from having run up the hill from the creek. Uncle Jim was also flushed white and shaking with fear. He'd been snake bit.

Daddy naturally asked, "What kind of snake was it?"

"It got away," said Uncle Jim. "It could've been a cotton-mouth."

Cotton-mouth snakes are more typically seen in swampy lakes and are rare along creeks, and Daddy knew it, but sometime copperheads come up the creek.

Mama said, "Show me the bite."

Uncle Jim held out his hand. Sure enough, there were a couple of small punctures in the skin of his wrist that could possibly have been the result of a snake.

Daddy asked, "How big was it?"

Lyle said, "Not too big."

Uncle Jim said, "It was big enough to bite me!"

Mama said, "William, I think you should suck out the poison."

Daddy didn't look like he was too eager to go sucking on Uncle Jim's hand. He said, "Call the doctor and see what he says."

Mama got on the phone. Dr. McCready was evidently not close to the phone because Mama had to sit there holding the phone to her ear doing nothing but look fretful while the rest of us waited. We all waited except Daddy. He had gone to the bathroom to get alcohol and a single-edged razor to get ready to do his bit to save Uncle Jim's life.

About the time Daddy got back and was getting ready to start cutting on Uncle Jim, Mama started talking. "Dr. McCready, this is Carolyn Gill. Jimmy is here and he's been bitten by a snake."

She listened to Dr. McCready's question and relayed it to Uncle Jim. "How long has it been?"

"We ran straight here. It can't have been more than twenty minutes ago."

She told the doctor and listened to the next question. She looked at the bitten hand and said, "No, it doesn't look swollen."

Another wait, and her answer this time was, "The bites don't look too deep. In fact, you have to look close to see them."

Then as she listened, she smiled. We could tell she was relieved, but she still asked one more question. "Do you think William should try to suck out the poison?"

After she hung up, she said, "Dr. McCready said that if you'd been bitten by a poisonous snake, you would be dead about halfway up the hill."

Uncle Jim clearly didn't like this diagnosis, but began to come around when Mama added, "He also said there was bound to be a lot of swelling by now if it was a poison snake."

"He didn't think it should be sucked?"

"No! He said he'd had to treat too many infected cuts from people who thought they'd been snake bit and made the cuts too deep and slobbered all over it. He said leave it alone."

Uncle Jim recovered quickly. "I guess we still have time to do gigging." Lyle agreed, and the two headed back to the creek.

Swimming tired

In recent years, much of the toil and sweat involved in making hay has disappeared. It used to be a time for all the kinfolk and anybody else who could be found to make the bales then stack them on a truck or trailer for hauling into the barn. Then came the arduous task of getting the hay into the barn and carefully stacking so the resulting pile would remain strong and stable until the hay was needed next winter. The work was hot and dusty, and it took us many hours to complete. After a day of working in hay, we were ready to drop from exhaustion and the dirt and sweat were ground into eyelids and hair and ears and elbows and every part of every one of us.

There are few people in the world dirtier or more worn out than hay workers after a hot June hay day. They may act like they never want to move again, until it dawns on somebody that the creek is near, and it is running clean and clear and is cool and very, very wet. Tired muscles stir. Everybody, young and old, starts thinking about the welcome cool shade and the beckoning creek water. Before another minute has passed, someone starts the truck, and everyone piles in the back. Shoes come off. Boys with shirts peel 'em off. By the time the truck gets to the creek, everybody is ready to hit the water hard and fast.

Nothing in the world feels better than working hard, getting filthy and sweaty, then swimming in the creek. First thing is to go under and get rid of as much dust and sweat as possible. Then there is the obligatory horse play. Muscles that were practically devoid of energy only moments before came to life and found a way to propel a mischievous kid to cannon ball into the midst of the crowd, wetting everyone, even the hapless ones who'd made the ill-fated decision that they would cool off with a little wading.

Tired people don't play as long and hard as children. It is late in the day, and the creek is cool and fresh. The water comforts the weary. The creek is alive with life. Fish swim in water -- clear or muddy -- fresh or scummy. There are many little fish and a few old ones, wise in the ways of staying alive. Turtles lurk in hidden places or crawl to gather sun upon their backs. Insects fly above the water or skate on its surface. Trees cool the water with their shade as they suck the moisture into their leaves.

The last lone swimmer simply floats on his back, nothing above the surface except his face. He barely moves. Minnows gently nose his skin, exploring this strange giant intruder. Creek water is good at carrying sound. He can hear his own breathing and heartbeat. There's the creek gently slapping the bank, and the trickly sound of water falling into the quiet pool.

Snake doctors fly low, near the surface. AKA dragonflies, snake doctors often fly two together, head to tail. Maybe mating – maybe just flying together. Gnats swarm in a couple of places, mostly near the shore.

A Cane Creek swimming hole

The swimmer's eyes are closed, and the current slowly takes him down the quiet stream. He opens his eyes, but the sun blinking through overhanging leaves tells him he has time, plenty of time. He closes his eyes again. The creek has always been and will always be. Snakes will not bother him. There is nothing for the resting swimmer to worry about. He can simply drift. There is time.

CPSIA information can be obtained
at www.ICGtesting.com
Printed in the USA
LVHW070807110122
708269LV00010B/107